Controlled Release Systems: Fabrication Technology

Volume II

Editor

Dean Hsieh, Ph.D
President
Conrex Pharmaceutical Corp.
Brandamore, Pennsylvania

CRC Press, Inc.
Boca Raton, Florida

Library of Congress Cataloging-in-Publication Data

Controlled release systems.

  Bibliography: p.
  Includes indexes.
  1. Drugs--Controlled release. 2. Controlled release
preparations. I. Hsieh, Dean, 1948-   .
RS201.C64C665   1988   615'.19   87-21785
ISBN 0-8493-6013-7 (v. 1)
ISBN 0-8493-6014-5 (v. 2)

   This book represents information obtained from authentic and highly regarded sources. Reprinted material is quoted with permission, and sources are indicated. A wide variety of references are listed. Every reasonable effort has been made to give reliable data and information, but the author and the publisher cannot assume responsibility for the validity of all materials or for the consequences of their use.

   Direct all inquiries to CRC Press, Inc., 2000 Corporate Blvd., N.W., Boca Raton, Florida, 33431.

International Standard Book Number 0-8493-6013-7 (volume I)
International Standard Book Number 0-8493-6014-5 (volume II)
Library of Congress Card Number 87-21785
Printed in the United States

FOREWORD

Controlled release technology is a rapidly emerging field. Two decades ago, formulations based on this technology were scarcely in existence. Today, the number of controlled release products is large and growing rapidly. With the advent of biotechnological advances and genetic engineering, the development of new and more complex drugs is imminent. These developments have necessitated more than ever before the creation of effective delivery systems which can protect these precious molecules from destruction, yet continuously deliver them to the body safely. Central to the successful development of any controlled release system is the fabrication of formulation procedure — it must be safe, reproducible, not damaging to the drug, and amenable to scale-up. The choices as to the types of controlled release systems one might use and the different ways these systems can be fabricated are significant. Such choices can only be made once an understanding of the principles underlying these systems and fabrication procedures is in hand.

While many books have been written on controlled release, this book stands alone in its effort to bring together an understanding of the ways in which controlled release systems can be fabricated. Dr. Hsieh is certainly well qualified to edit and put together such a book, as he is a skilled formulator of controlled release systems and has studied them extensively both as a postdoctoral scientist in our laboratory at MIT, as a Professor at Rutgers, and at his company, Conrex Pharmaceutical Corporation. This book brings something new to readers in the exciting field of controlled release technology/fabrication technology, an area that should be of practical laboratory value in the design of these important systems.

Robert Langer

PREFACE

Controlled release technology has well deserved its recent growth in popularity and widespread acclaim. Its advantages have been recognized and utilized not only by the pharmaceutical industry, but by several other industries as well. Advances in controlled release research enable innovative refinements in many currently manufactured household products. Employment of these refinements helps the sponsors of such research to achieve a competitive edge. However, before controlled release products can be manufactured, further research and development must take place. Fabrication technology refers to the methods by which controlled release products are manufactured. Throughout the development of controlled release devices, fabrication technologies have played a key role in the process of innovation. The purpose of these volumes is to compile and generalize principles of fabrication methods which have been previously published. These volumes thus provide a framework for the study of fabrication technology. It is the editor's hope that they will form the basis for future innovation.

The first volume is concerned with fabrication procedures for currently marketed products or mature technologies. The second volume is concerned with fabrication procedures in various stages of development. Some of the technologies described in the second volume may be mature, yet may belong to a class of products wherein the majority are still under development. Volume I, Chapter 1 is concerned with this classification and explores it in detail. Briefly, there are three stages in the development of controlled release technology:

1. Encapsulation technologies are covered in Chapters 2 through 7. These fabrication procedures have matured, now comprising, for instance, coacervation, film coating, and mechanical blending.
2. Transdermals and other advanced drug delivery systems are covered in Chapter 8 and Volume II, Chapters 1 through 7. These include multiple lamination for transdermal patches, injection molding, extrusion, gelation, multiple emulsion, and other methods for fabricating bioerodible and hydrogel drug delivery systems.
3. Selective drug targeting systems are surveyed in Chapter 8. These systems represent the latest and most promising stage in controlled release technologies. They include monoclonal antibodies, liposome delivery systems, dextran and magnetic microspheres, and polymeric delivery systems.

All three stages in controlled release technology must utilize various sterilization procedures according to individual fabrication processes. Chapter 9 provides an overview of sterilization procedures for controlled release products.

Much of the knowledge contained within these volumes will prove valuable to the scientific community. For this reason, it has been difficult for contributors to disclose proprietary information. The organizations and individuals who have offered the results of costly and painstaking research deserve greater rewards than I can offer. Nevertheless, I take this opportunity to thank all the contributors for the time and effort they have devoted to this project. I also thank those friends whose guidance directed me to these outstanding individuals. These volumes are truly the product of a concerted effort by a talented and enthusiastic group of colleagues. Those with expertise and a willingness to share it have made this book possible to complete. Even those authors whose companies vetoed their contributing chapters added stimulus to the project with their initial enthusiasm.

In addition to the major contributors to these volumes, there have been several individuals whose continuous support has sustained me through the project. I thank my wife, Mrs. Phyllis Hsieh, for her unqualified patience, endurance, and encouragement. I also thank J. C. Lorber for the capable and prudent input as well as the cooperative effort necessary to make constant and steady progress. I would also like to acknowledge the assistance of my colleagues at the Rutgers College of Pharmacy: Drs. Y. Chien, K. Tojo, and C. Liu, among others. Their criticism concerning the pursuit of this project, both positive and negative, has been appreciated. This appreciation extends to several graduate students at Rutgers, including P. Mason, C. C. Chiang, E. Tan, and R. Bogner, who helped to proofread the communications involved in the project. Furthermore, I must thank the editorial staff at CRC Press, Inc., for the experienced coordination of communication which led to final agreements with the contributing authors. Finally, my thanks to CRC Press for giving me this opportunity to pursue this most significant and greatly rewarding project.

Dean Hsieh

EDITOR

Dr. Dean Hsieh is the founder and president of Conrex Pharmaceutical Corporation, Brandamore, Pennsylvania since 1985. He obtained his Ph.D. ('78) and M.S. ('74) degrees from M.I.T., followed by postdoctoral training with Professor Langer at the Boston Children's Hospital, associated with the Harvard Medical School. In January, 1981, he was promoted to Instructor at Harvard Medical School. In October 1982, he became an Assistant Professor at the College of Pharmacy, Rutgers — The State University of New Jersey. He has authored and co-authored more than 60 papers and abstracts in the area of drug delivery systems and controlled release technologies. He is also the holder of several patents and pending patents. Recently, his efforts have focused on the research and development of proprietary permeation enhancers, leading to the commercialization of this technology.

CONTRIBUTORS

J. Chen
Agricultural Division
Ciba-Geigy Corporation
Greensboro, North Carolina

Diana Chow
Student
Department of Pharmaceuticals
Texas Medical Center
University of Houston
Houston, Texas

Gary Cueman, M.S.
Department of Pharmaceutics
Rutgers College of Pharmacy
Piscataway, New Jersey

Thomas H. Ferguson, Ph.D.
Research Scientist
Animal Health Formulations
Eli Lilly & Company
Greenfield, Indiana

Dean S. T. Hsieh, Ph.D.
President
Conrex Pharmaceutical Corporation
Brandamore, Pennsylvania

Joachim Kohn
Assistant Professor
Department of Chemistry
The State University of Rutgers
Piscataway, New Jersey

Robert Langer, Sc.D.
Professor of Biochemical Engineering
Department of Applied Biological
 Sciences
Massachusetts Institute of Technology
Cambridge, Massachusetts

Ping I. Lee, Ph.D.
Director
Basic Pharmaceutics Research
Ciba-Geigy Corporation
Ardsley, New York

Kam Leong
Assistant Professor
Department of Biomedical Engineering
Johns Hopkins University
Baltimore, Maryland

Virgil L. Metevia
Technical Manager
Dow Corning STI
Walnut, California

Richard C. Oppenheim, Ph.D.
Senior Lecturer
Department of Pharmaceutics
Victorian College of Pharmacy Limited
Parkville, Australia

Howard B. Rosen
Engenics, Inc.
Menlo Park, California

W. Eugene Skiens, Ph.D.
Manager, Media Development
Optical Data Inc.
Beaverton, Oregon

Patrick A. Walters, M.S.
Technical Manager
Medical Materials Technical Service and
 Development
Dow Corning Corporation
Midland, Michigan

David Woodford, Ph.D.
Assistant Professor of Pharmaceutics
Pharmaceutics Department
College of Pharmacy
University of Rhode Island
Kingston, Rhode Island

Joel Zatz, Ph.D.
Professor
Department of Pharmaceutics
Rutgers College of Pharmacy
Piscataway, New Jersey

TABLE OF CONTENTS

Volume I

Volume II

Chapter 1

NANOPARTICLES: SOLID SUBMICRON DRUG DELIVERY SYSTEMS

Richard C. Oppenheim

TABLE OF CONTENTS

I. INTRODUCTION

When a drug delivery system is fabricated with controlled release, it is normal that the mechanism and rate of drug release from the system have been identified and built into the product. However, there is usually a poor definition of or control over the site of release in relationship to the actual or perceived site of action.

Most controlled release products are developed to bypass some inconvenient physiological situation (e.g., those created by enteric coated products) or to have some form of sustained or extended release profile. In neither case is there any real attempt to ensure that the released drug encounters an appropriate site of action. Provided that the circulating concentration is sufficient to ensure the correct response at the desired site of action and that any responses at undesired sites of action are not life threatening, then such a controlled release system is adequate. It does, however, represent poor application of modern pharmaceutics.

With the current regulatory difficulties in getting new but not revolutionary drugs marketed, pharmaceutical companies are developing new delivery systems for approved drugs. In general terms, the site of action is known. A reasonable estimate of the concentration of the drug in the immediate neighborhood of that site, if not at the site itself, can be developed knowing the pharmacokinetic and pharmacodynamic properties of the drug. Hence, there is often sufficient information available to address questions in the second area of controlled release.

This chapter will focus on those submicron particulate delivery systems in which a deliberate atttempt is made to respond to the problems in both of these areas. Hence, systems which cause uncontrolled or serendipitous therapeutic responses will not be discussed. Examples of such excluded particulate systems are inhaled cigarette smoke, metal or asbestos dusts, viruses or pollen, contaminating silica particles leached off blood defoaming bags used in open heart surgery, or particulate contamination derived from the administration of large- or small-volume parenterals. Similarly, simple diagnostic and therapeutic colloidal systems such as technetium-labeled colloidal sulfur or colloidal gold as [198]Au have been discussed previously.[1]

Widder et al.[2] clearly identified the three levels of targeting at which modern parenteral drug delivery systems should be aimed: (1) a local perfusion of an organ by drug released from the delivery system, (2) interaction of the released drug only with part of an organ or even with the external membrane of the target cells, and (3) no release of drug until the delivery system is inside the cell containing the site of action. Conceptually the fourth and ultimate level is delivery to and release at the site of action. As this targeting approach to drug delivery is more widely accepted and achieved, there will be a much reduced need for biopharmaceutics as it is now practiced.

Tomlinson[3] has modified this approach by dividing targeting into two types, passive and active. In passive targeting, the size of the particulate delivery system is used to define the type of automatic response by the body to that delivery system. Capillary blockage is possible for particles greater than approximately 4 μm in diameter. If the particle diameter is between 30 nm and 7 μm, there can be reticuloendothelial system (RES) uptake and general lysosomotropism, whereas if the particles are less than 100 to 200 nm in diameter, they can pass through basement membranes and escape from the circulatory systems.

In active targeting, which seems to be useful for particles with a size of 0.05 to 4 μm, the formulation and/or the user guides the particle in some way as it interacts with the body. This guiding procedure could be antibody attachment to the particle so that specific cell antigens can be identified. Alternatively, magnetite could be incorporated into the particle and an external magnet focused on the target part of the circulatory system, so that as the particles arrive at that part, they are trapped there.

There are a number of consequences[1] of controlling the release of a drug by targeting it in some way. The circulatory levels of free and protein-bound drugs are reduced drastically. The total amount of drug used is reduced. The cost of the delivery increases due to the technology of making the delivery system. Provided that this increased cost is not as great as the savings achieved in reduced drug dosage, reduced hospital residence time, and reduced repair of adverse and toxic reactions to the drug, the targeted controlled release system could be economically viable.

A variety of terms have been used to describe the solid particulate systems used to achieve controlled parenteral release of a drug. Tomlinson[3] describes all particles with a size of between 20 nm and 300 μm as microspheres. While this has the advantage of simplicity, it is clear that capillary blockage cannot be achieved by submicron- or nanometer-sized particles. Consequently Oppenheim[1] and Kreuter[4] have argued that a functional definition based on size of the particle is appropriate. Hence nanoparticles are solid colloidal particles ranging in size from 10 to 1000 nm where the active ingredient is dissolved, entrapped, encapsulated, adsorbed, or attached. It follows that nanoparticles have a direct interaction with individual target cells, i.e., either the second or third level of targeting.[2] Microspheres are similar particles, but with a size greater than about 4 μm, and as such can be embolized in capillaries. The released drug can then interact with the target cells. This corresponds to the first level of targeting.[2] It is suggested that particles in the overlap size range of 1 to 7 μm be classified on the basis of their embolism (microspheres) or alternatively their RES interaction or active targeting approach (nanoparticles).

It is vital to differentiate these two fundamentally different approaches to drug delivery because each requires a different approach to controlling the release of the drug. Microsphere delivery involves the payload of drug being released by leaching from pores, diffusing through an external coat, or becoming available as the microsphere slowly degrades. A small release of drug prior to embolization is not desirable, but is not disastrous from a formulation point of view. On the other hand, nanoparticle delivery can only be at the surface of the target cell or, preferably within the target cell. Hence there should be no leakage prior to this target dumping. This stringent control of release is difficult to achieve. It also leads to a dilemma for the system designer. The desired intracellular release is controlled only by the amount and character of the intracellular, often lysosomal, enzymes of the target cell over which the designer has no control. Therefore, an increase in targeting efficiency can lead to a loss of control over release rate.

Many more microsphere systems than nanoparticle systems have been developed. In general terms, this is because they are simpler to make and use. A more complete detailed compilation of the properties and behavior of the more important microsphere delivery systems is available in recent books edited by Illum and Davis[5] and Davis et al.[6] In this review of nanoparticles, the relevant key parts of the information available about the manufacture, storage, and use of microspheres will be included. A rational approach will be developed for fabricating and fomulating nanoparticle delivery systems which will be able to achieve upper-level targeting.

II. SYSTEM LIMITATIONS

Any drug delivery system is a combination of the active drug and the base carrier molecules. Physical, chemical, and physiological properties of each part of the system limit the choice and the combination of choices for each part. As nanoparticles will deliver the drug to the membrane or to the intracellular fluid of the target cell, the system designer must have information about the interaction mechanism and strength between the drug and the external membrane; the partition coefficient of the drug between the intracellular fluid, the cell membrane, and the surrounding fluid; and the drug concentration required either in the membrane or in the intracellular fluid for effective therapy; as well as the metabolic pattern of the drug associated with that target cell.

For parenteral administration of particles, the individual as well as the associated, aggregated, or polymerized carrier molecules must be nontoxic, nonimmunogenic, and biodegradable. Acceptance of these criteria severely limits the choices available to the system designer. Many systems have been developed and tested in which a compromise has been reached with these criteria.

The combination of the drug and the carrier molecules used to make the drug delivery system must meet a number of criteria. The system must be pharmaceutically acceptable with regard to shelf-life. Hence most of the vast literature on liposomes does not help formulate better solid submicron drug delivery systems.[7] The system must be sterile before administration: filtration cannot be used since the nanoparticles are similar in size to the contamination particles. Additionally the system should be easy to administer with standard equipment without causing any localized reaction. The system should not cause any hapten production or other immunological reaction; as the carrier molecules are not immunogenic, any such problem would be the result of the drug and the carrier molecules interacting. The delivery system must have an acceptably large payload of drug. The system must be designed so that any opsonization that takes place after administration either facilitates or at least does not hinder the targeting. The system must release its payload of drug only when the system can identify the site of action of the drug. Release must then occur at a predetermined controlled rate by the release mechanism built into the system. Much work has been undertaken on release mechanisms and rates without due regard to all the other system criteria. Finally, if there are parallel shifts in the LD_{50} and the minimum effective dose, then the therapeutic index of the drug in the particulate and free form is the same. Similarly, the particulate delivery system could increase the concentration of the drug in the target cell and in important nontarget cells equally.[8] In neither case does the controlled release product offer any advantage.

These system limitations are severe and have not been totally overcome. Consequently nanoparticle manufacture is generally limited to small-scale, noncommercial batches. Scale-up problems will be encountered, but as considerable experience has been accumulated with microspheres, scale-up is not expected to be a major difficulty with nanoparticles. When the first types of nanoparticles (those based on the acrylates) are discussed, the approach adopted to try to solve the extra- and intrabody questions posed at the beginning of this chapter will be highlighted. As each of the other types of nanoparticles is discussed, mention will be made of these approaches, particularly as they influence manufacturing, storage, and administration procedures. Detailed discussion of the use to which these controlled release nanoparticles are being put, experimentally and clinically, is not given here, but is available in the references cited.

III. ACRYLATE SYSTEMS

There are a number of base molecules which may be used and their choice is determined by how the nanoparticle is to be used. Kreuter[9] recommends the almost nonbiodegradable poly(methylmethacrylate) nanoparticles for vaccination purposes because the antigen is retained for a long time, leading to prolonged immunostimulation. Similarly, enzymes[10] and proteins[11] have been immobilized in polyacrylamide microbeads and nanoparticles to ensure long-term activity in industrial processes. On the other hand, polycyanoacrylate nanoparticles are more suitable for use as parenterally administered drug carriers because they have a controllable, quicker biodegradation rate.[9]

All polyacrylic particles are made by techniques based on emulsion, suspension, or bead polymerization. Emulsion polymerization, either with or without added surfactant, generally leads to submicron particles. Suspension polymerization yields microspheres while the beads formed by the third technique usually have millimeter-sized diameters.

Despite the fact that these techniques are used on a large scale by a variety of industries, polyacrylic particles are still only produced in small batches for experimental therapeutic use. However, there seem to be no significant scale-up problems apparent, once a drug-containing product has been shown to release its payload at a controlled rate by a known mechanism at the desired physiological site. Many potential scale-up problems have been encountered and resolved in the enzyme immobilization industry. Whether the scale-up would lead to an economically viable product depends on the balance of manufacturing costs and health care savings. The particles can be made in either organic or aqueous phase using polymerization procedures described in the following sections.

A. Manufacture in Organic Phase

Initially[12] Sjöholm's group in Sweden used 200 mℓ of a 4:1 toluene:chloroform organic phase into which 5 mℓ of a buffered aqueous solution of acrylamide, drug or active macromolecule, and ammonium peroxodisulfate was emulsified using a high-speed homogenizer and 0.25% Pluronic F68 as emulgent. The composition of the organic phase was chosen to give a density of 0.98 g/mℓ in order to facilitate a stable emulsion. N,N^1-Methylenebisacrylamide can be added to the aqueous phase when a reduced biodegradation rate and a more porous particle is required. After adding N,N,N^1,N^1-tetramethylethylenediamine as the initiator, radical polymerization proceeded with gentle stirring at room temperature to produce 1000-nm nanoparticles. Oxygen had to oe excluded to ensure successful polymerization. Ekman et al.[13] had found earlier that high-speed homogenization during the polymerization led to 1- to 10-μm particles, presumably due to collisional coalescence of the polymerizing particles.

Edman et al.[14] used much the same manufacturing technique to make nanoparticles based on molecules which had dextran or other polysaccharides incorporated into the acrylic monomer. Large molecular weight materials such as albumins or enzymes can be immobilized in each type of system partly by entrapment in the polymeric network and partly by physical fixation within the polymeric bundles. The polysaccharide particles immobilize a greater payload of protein than the polyacrylamide particles because as the pore size is reduced in the polysaccharide system, the extent of entrapment increases. Albumin payloads of 35 to 40% can be achieved[15] in the polysaccharide particles. However, although the salicylate binding capacity of albumin polyacrylamide nanoparticles is not impaired, it is reduced by 50% when used in maltodextrin particles. These polysaccharide particles also leach their payload more rapidly than the polyacrylamide particles, presumably due to a much faster degradation of the starch component of the particles in both in vitro and in vivo situations.[16] Hence, a balance of the starch component and the hydrocarbon chain is essential to ensure an optimal release profile in vivo.

Low molecular weight drugs cannot be effectively entrapped in either type of acrylamide-based nanoparticle because they readily leach from the pores. El-Samaligy and Rohdewald[17] found similar results for 100 to 600-μm polyacrylamide microbeads. However, it may be possible to conjugate such drugs to the glucose residues, make the nanoparticles, and then use lysosomal enzymes to cleave the drug-glucose link and release the drug. Assuming that such a conjugation does not enhance the degradation of the nanoparticle before it encounters its target cell and that this type of nanoparticle can be taken up by that cell, the release rate is determined only by the amount and character of the lysosomal enzymes in the target cells and not by any feature built into the delivery system by its designer. The pore size of these nanoparticles is too great to be of use to the designer in controlling the drug release. Considerable work remains to be done to control the pore size of these and all other porous nanoparticles. If this can be achieved, then pore diffusion could become a rate-limiting step in controlled drug delivery from nanoparticles.

Intravenous and intraperitoneal injections of L-asparaginase-containing 250-nm polyacrylamide nanoparticles in mice produce an enhancement of the humoral antibody response compared to a soluble enzyme given by the same route.[18] Intramuscular or subcutaneous injection of the particles gave a much weaker immune response. However, the ability of the particle-entrapped enzyme to react with an antibody is decreased compared to the soluble enzyme.

Edman et al.[19] found that with 40-mg/kg body weight intravenous single doses of 400-nm polyacrylamide nanoparticles in mice, only insignificant and hardly detectable morphological reactions could be seen in the liver, spleen, and bone marrow. Doses of 160 mg/kg were needed to produce significant adverse reactions. If such a system is to be used as a drug delivery system, this then sets the upper limit on the amount of nanoparticles and hence attendant drug payload that can be administered. No chronic toxicity studies appear to have been reported.

Similar polyacrylic nanoparticles have been developed independently by Speiser's group in Zurich. Birrenbach and Speiser[20] used hexane as the outer phase. By using 15% of *bis*-(2-ethylhexyl)-sodium sulfosuccinate and 7.5% of polyoxyethylene-4-lauryl ether, up to 35 mℓ of water can be solublized in 80 mℓ of hexane. The acrylic monomers such as acrylamide or methylmethacrylate readily dissolved to form a clear solubilized system. The polymerization was preferably done by gamma irradation in a "cobalt bomb" (300 krad) or by UV irradiation. It could also be achieved by visible light irradiation provided small amounts of a catalyst such as riboflavin-5′-sodium phosphate or potassium peroxodisulfate had been added. This latter method produced fewer polydisperse nanoparticles than the gamma irradiation method.[21] Much smaller 40-nm water-swollen polyacrylamide particles stabilized by Aerosol OT and dispersed in toluene have been made and characterized by Candau et al.[22]

Kreuter[4] argues that the lower solubility of the acrylic monomers in the hexane compared to the toluene-chloroform mixture used by Sjöholm's group leads to a reduction in the extent of acrylic monomer molecule diffusion. This results in an increase in the stability of the system which means stirring is not required during the polymerization. This could be the reason for a greater batch-to-batch and within-batch uniformity of particle size in the systems using hexane. The problems with all these acrylamide-type systems are poor biodegradability, concern that all the unincorporated monomer is removed in the phase separation/centrifugation cleanup procedure, and the possible degradation or denaturation of the active payload upon irradiation.[23]

Kreuter[9] found that scanning electron microscopy (SEM) (after allowing for the thickness of the conducting gold coating), transmission electron microscopy (after freeze fracturing), and photon correlation spectrometry gave comparable estimates of size for both poly(methylmethacrylate) and polyacrylamide nanoparticles. Using nonadsorbing helium in a gas pycnometer, he found the density of polyacrylamide and poly(methylmethacrylate) nanoparticles to be 1.14 and 1.06 g/cm^3, respectively. This latter figure is consistent with the porous nature of the nanoparticles when compared to 10-μm beads of the same material which had a density of 1.15 g/cm^3. All the polyacrylic nanoparticles were X-ray amorphous. The electrophoretic mobility of aggregates of 120-nm poly(methylmethacrylate) nanoparticles was about -1.3 μm.cm/(sec.V) in an unknown-strength phosphate-buffered saline at pH 7.4. This mobility dropped to about -0.25 μm.cm/(sec.V) when the nanoparticles were stored in human serum, indicating opsonization of the particles. Polyacrylamide nanoparticles did not readily aggregate and no electrophoretic measurements were made. No zeta potential or surface charge calculations are possible on these systems which have aggregated to an unknown extent. This opsonization process is reflected in the drop in contact angle of water on compressed particles from 73 to 53° before and after storage for 12 hr in human serum.

B. Manufacture in Aqueous Phase

Alkylcyanoacrylates are practically insoluble in water. The alkyl groups commonly found in the polymerized form of these acrylic materials used as biodegradable or dissolvable sutures are methyl, ethyl, propyl, and butyl. Nanoparticles can be made from these materials by solubilizing the monomer in a surfactant containing aqueous vehicle and allowing anionic polymerization to take place. Again very small batch sizes are used with, for example,[24] 0.25 mℓ of butylcyanoacrylate being added dropwise to 24.75 mℓ of a 0.2-μm filtered, stirred aqueous solution of 0.5% w/v Dextran 70 in 0.01 M hydrochloric acid at 20°C. A magnetic stirrer set at about 1000 rpm is used to fully disperse the monomer. After 2 hr, the polymerization is complete and the product can be cleaned up by filtration, centrifugation, or dialysis. Other surfactants which have been used include Polysorbate 20,[25] Pluronic L63,[26] and Pluronic F68[27] (both polyoxyethylene-polyoxypropylene surfactants). Illum et al.[28] used the less hydrophilic material Dextran 70 when the surface of the nanoparticle was to be the adsorption site for the Fc portion of antibodies, leaving the F(ab) portion accessible to recognize cell membrane antigens.

In a detailed study, Douglas et al.[24] found that control of particle diameter in the range of 100 to 200 nm could be achieved by varying the pH of the system between 2 and 3. At pH values below 2, the rate of the anionic polymerization is too slow and a very polydisperse system results due to the coalescence of nanoparticles swollen with unreacted monomer. A higher pH values, the increased hydroxyl ion concentration leads to a very rapid polymerization rate and an amorphous polymer mass results.

At monomer concentration below 1%, the nanoparticles were quite polydisperse. Reasonable control of the size and dispersity was achieved at monomer concentrations between 1 and 2.5%, but at higher concentrations up to the maximum of around 7% (when the resultant system was not free flowing) the particle size increased markedly and the dispersity increased slowly. Other physicochemical factors such as temperature and acidifying agent had no effect on particle size. Since the added electrolyte over the concentration range of 0.01 to 0.25 mol/dm^3 also had no effect on particle size, surface charge of the particle is not important with respect to particle size or maintenance of colloidal stability, inferring that the system is sterically stabilized. This is in direct contrast to the charged polystyrene latices which are highly ionic-strength dependent.[29] Douglas et al.[24] also found that stirring rate had no effect on particle size, although Kreuter[4] suggests that vigorous stirring can lead to larger butylcyanoacrylate particles. In a second report, Douglas et al.[30] found that the diameter of the nanoparticles could be controlled in the range of 20 to 770 nm by varying the molecular weight and concentration of dextrans, poloxamers, and polysorbates.

Kreuter[9] also investigated the physicochemical properties of polyalkylcyanoacrylate nanoparticles. He found that the presence of surfactants used in their manufacture led to a smooth coat over the particle surfaces, which inhibits the observation of the individual structures. Removal of the surfactants, however, often results in aggregation and leads to irreversible changes in the nanoparticles. Hence SEM is not very useful for these products, nor can it determine the fate of the surfactant when the nanoparticles are administered in vivo or whether or not the nanoparticles can remain discrete in vivo if the surfactant desorbs.

Removal of the surfactant caused nanoparticle aggregation which enabled electrophoretic mobilities to be determined. A clear chain-length effect was found in the phosphate-buffered saline at pH 7.4 in that the methyl, ethyl, and butyl derivatives had mobilities of -1.64, -1.32, and -0.87 μm.cm/(sec.V) respectively. This effect was completely obliterated when the measurements were done in human serum when a value of around -0.22 μm.cm/(sec.V) was determined. This is about the same as the mobility of the poly(methylmethacrylate) nanoparticles, indicating that the effect on the surface after opsonization was very similar for the two types of acrylic nanoparticles. The water contact angle measurement indicated that the butyl derivative was more hydrophobic than the methyl derivative. This marked hydrophobic effect was lost upon storage in human serum, and again there was little difference from the poly(methylmethacrylate) nanoparticles stored in human serum.

As a carrier and controlled deliverer of drugs, polyalkylcyanoacrylate nanoparticles offer a number of features. First, unlike the acrylamide-based nanoparticles, polyalkylcyano-acrylate systems are biodegradable. Chemically this involves hydrolysis of the carbon chain, forming formaldehyde and an alkylcyanoacetate. The rate decreases as the alkyl chain length increases.[31] Kante et al.[32] found that although 0.5% polyisobutylcyanoacrylate nanoparticles could be incubated in vitro with hepatocytes without increasing cell mortality, a particle concentration of 1% caused marked cell death. This could be due to the formaldehyde products of chemical degradation of the nanoparticle. The methyl product degrades faster than the isobutyl product and is much more toxic. This intracellular effect of formaldehyde is analogous to that found in gelatin implants[33] and proteinaceous nanoparticles (see Section IV.A). However, Couvreur et al.[34] show that isobutanol can be produced from this type of nanoparticle incubated with rat liver microsomes. It is possible that the enzymatic degradation pathway is different from the chemical pathway and occurs at a faster rate, but can be inhibited by one of the degradation products. Much more work is required to define the degradation pathway, its kinetics, and the inhibitory or toxic degradation product. Hence, although the alkylcyanoacrylate nanoparticles are biodegradable, this cannot occur at too rapid a rate. Therefore, particle degradation cannot be used as a means of controlling the release of the payload of the drug unless the concomitant release of formaldehyde or other inhibitory degradation products assists the drug in causing the desired death of the cell. However, such a rapid particle degradation may cause the nanoparticle to be degraded in vivo before it reaches, interacts with, or enters the target cell.

This has to be investigated for each system being developed. Brasseur et al.[35] showed that injection of 3 mℓ/kg of body weight of rats of a 0.5% w/v polymethylcyanoacrylate nanoparticle solution caused no deaths nor any inhibitory action on implanted soft tissue sarcoma S250. For such particles with a payload of 1.3% actinomycin D, there was both a reduction in tumor size and an increase in animal death which could indicate some targeted controlled release, but also some premature nanoparticle degradation. It is interesting to note that Illum et al. have recently used polyhexylcyanoacrylate nanoparticles in vitro[28] and in vivo.[27] In principle, different types of alkylcyanoacrylate nanoparticles could be blended together[36] to give any desired release profile just as is traditionally done using microcapsules with different wall thicknesses.

The second feature of alkylcyanoacrylate nanoparticles as a system to deliver drugs in a controlled way to target cells is the extent that and the way in which the payload of drug can be incorporated into the nanoparticle. Typical payloads have been given by Couvreur[36] for a range of cytotoxics as dactinomycin (0.3 to 0.7%), vincristine (0.8 to 2.0%), vinblastine (3.0 to 5.5%), methotrexate (0.2%) and doxorubicin (9.2%). As the ratio of drug to monomer increases, the percentage of drug in the overall system incorporated into the forming nanoparticles decreases, but the payload of drug in the nanoparticle increases.

Active material can be incorporated as the nanoparticle is being made. Kreuter et al.[37] incorporated [^{75}Se] norcholestenol into polybutylcyanoacrylate nanoparticles. About 1/6 of the payload can be rapidly (1 day) dialyzed into phosphate-buffered saline, the balance being released at about 1/15 the rate. The release rate into horse serum was about 10% of that in the buffered saline. The porous nature of the nanoparticles facilitates the rapid initial dumping of the payload and necessitates the storage of the product as a lyophilized powder. It also precludes the nanoparticle from delivering 100% of its payload to the cells to which the nanoparticles target.

El-Egakey and Speiser[25] showed that verapamil hydrochloride could sorb onto preformed 400-nm polybutylcyanoacrylate nanoparticles. The resultant Langmuir sorption isotherms indicated that as the drug pK_a was approached, more efficient sorption took place. An increase in electrolyte strength increased sorption, probably by a salting-out effect, whereas increased amounts of surfactants such as Polysorbate 20 or Pluronic F68 decreased sorption either by blockage of surface sites or by drug solubilization in the vehicle. Temperature studies yielded a small value of -11.8 kJ/mol for the standard enthalpy of sorption.

From these two studies it is clear that the payload of drug in porous polyalkylcyanoacrylate nanoparticles is partly made up of readily released drug and partly of drug which is either strongly bound or entrapped within the particles. Extrapolation of the data of Couvreur[38] suggests that between 20 and 30% of the added actinomycin is well incorporated and the easily adsorbed (and hence desorbed) material could be seen as poorly incorporated.

The resolution of this problem would seem to be to chemically conjugate the drug to the monomer, and once the particle is made, the drug would be released either at the rate of cleavage of the conjugate bond and/or at the rate of nanoparticle degradation at the system designer's wish. This prodrug conjugation clearly cannot be associated with the double bond needed for the anionic polymerization. Hence resolution of payload problems would be independent of problems and control of the manufacturing process.

The third feature of polyalkylcyanoacrylate particles is their size and how this can be used in passive targeting. Isobutylcyanoacrylate (known as bucrylate by radiologists) is commonly introduced by means of a silastic calibrated-leak balloon into cerebral arteriovenous malformations.[39] The monomer solidifies very quickly into beads in the malformation, occluding part or all of the abnormal vascular network. Radioopaque materials can also be mixed with bucrylate[40] to help visualize the embolic bead as well as the downstream region. In principle these beads formed *in situ* could be used as a means of locally perfusing an organ or a defined part of the circulation over an extended period.

However, most of polyalkylcyanoacrylate products have been made in the submicron, nanoparticle range. Hence in passive targeting, it is hoped that the nanoparticle drug delivery system can access the target cell, that the target cell will take up the nanoparticle, and that the drug is released at an appropriate rate. The normal nontarget cell is the RES, in particular, the various cells of the liver. Much more work needs to be done to try to identify and then utilize features of target cells and nontarget cells to try to improve the passive targeting efficiency. The most obvious target cell type is that of cancer. Until late 1983, attempts to target nanoparticles to tumor cells were likely to provide little success. If the nanoparticles had a diameter of more than 200 nm, they were unlikely to be able to passively pass through the basement membrane of the circulatory system and access the tumor cell. Lenaerts et al.[41] found that the ratio of uptake for rat liver Kupffer, endothelial, and parenchymal cells was the same for 80- as for 215-nm polyisobutylcyanoacrylate nanoparticles, indicating that the 100-nm fenestrated lining of the liver sinusoids was not exerting a great influence.

There is no doubt that a wide range of cytotoxic agents can be incorporated into polyalkylcyanoacrylate nanoparticles,[3,35] but extensive studies of such systems with a diameter of 100 nm or less, made during 1984, had still to be reported by March 1985 when this chapter was submitted. When in vitro tests have been disappointing, it is not always clear if the target cells actually take up the nanoparticles. This is usually checked by a fluorescent[42] or a radioactive[43] tag on blank, drug-free nanoparticles.

Given the problems discussed earlier about the potential rapid release of low molecular weight drugs, the work of Couvreur et al.[44] with insulin is encouraging. They found that the nanoparticles were effective in prolonging the hypoglycemic action after subcutaneous administration in streptozotocin-diabetic rats from 8 hr in control animals to 20, 24, and 26 hr in the ethyl, isobutyl, and butyl examples of the polyalkylcyanoacrylate nanoparticles. The peak plasma immunoreactive insulin levels were correspondingly lower than control levels. It appears that the results are in accordance with the degradation rate of the polymers and demonstrate that the drug was not altered by its incorporation into nanoparticles. In this example, there is no attempt to target the delivery system, but rather to provide a prolonged therapeutic response.

Given this apparent difficulty to passively target polyalkylcyanoacrylate nanoparticles, a variety of active targeting procedures have been developed and this is the fourth feature of polyalkylcyanoacrylate nanoparticles. First, the nanoparticle drug delivery system should be administered into the same "compartment" as the site of action. In joint degenerative diseases, e.g., arthritis, intra-articular administration achieves this: nanoparticles are unlikely to cross the synovial membrane into the general circulation. It is known[45] that invading polymorphonuclear cells are involved in the enzyme pathway which leads to increased degeneration. Such cells also avidly take up nanoparticles. By combining these properties, several groups have used nanoparticles intra-articularly. El-Samaligy and Rohdewald[46] made 210-nm polymethylcyanoacrylate nanoparticles with a payload of 2.5% triamcinolone diacetate. The drug was released in vitro into phosphate buffer at 37°C with a first-order rate constant of about 0.1/hr, and all the drug was released within 22 hr, since at this point the nanoparticles were completely degraded. Hence, release is a combination of porous diffusion and particle degradation. In principle such a system could be used in vivo for proteinaceous nanoparticles, as described later.

Rather than use the traditional intravenous administration route, Grislain et al.[47] administered 250-nm [^{14}C] polyisobutylcyanoacrylate nanoparticles subcutaneously and intramuscularly. They found the radioactivity concentrated in the gut wall and not in the liver and spleen. Kreuter et al.[48] showed that fecal elimination was very important following subcutaneously administered polymethylmethacrylate nanoparticles.

The second way of enhancing targeting of this type of nanoparticle was developed by Ibrahim et al.[49] They used the Widder et al.[2] approach (see Section IV.A.5) of incorporating magnetite into 220-nm isobutylcyanoacrylate nanoparticles simply by dispersing the 10- to 50-nm magnetite particles in the aqueous phase before adding the monomer. By applying a magnet over the left kidney of mice, three times more nanoparticles could be trapped there than in the control right kidney. There was also a reduction in liver uptake compared to control mice not subject to the magnet treatment.

While this approach will rapidly concentrate the drug delivery system at a predefined part of the circulatory system, the third approach to improving targeting requires no external intervention once the product is administered. The targeting device is built onto the surface of the nanoparticle, as it is the external surface of the nanoparticle that must come into contact with and recognize the external membrane of the target cell. Illum and Davis[50] used naked and Pluronic F108 (poloxamer 338)-coated 50-nm polystyrene latex particles; 90% of the uncoated particles were cleared by the liver and spleen with a half-time of 50 sec. The coated particles were only taken up to an extent of about 40%, and much larger amounts were found in the lung, heart, carcass, and blood. The clearance of the coated nanoparticles from the blood was much slower than the uncoated nanoparticles. They ascribe these effects to the hydrophilic nature of the surfactant and the decreased adhesion of poloxamer-coated nanoparticles to all surfaces. The application of these findings to polyalkylcyanoacrylate and other types of nanoparticles is still to be done. The effect of different opsonins on the redirection of interaction caused by the surfactant is likely to be most important.

With the recent emergence of the genetic engineering industry, antibodies to a wide range of cell surface antigens are now readily available. Despite the generally unpromising results obtained by antibody-coated liposomes, a number of groups have associated antibodies with polyalkylcyanoacrylate nanoparticles. Illum et al.[28] incubated monoclonal antibodies with 170-nm polyhexylcyanoacrylate nanoparticles overnight in phosphate-buffered saline. The complex retained its immunospecificity for at least 4 days at 4°C. However, the system showed no marked targeting efficiency in vivo.[27] This could be due to incorrect or reversible adsorption orientation of the antibody on the nanoparticle surface, desorption of the antibody in vivo caused by more strongly adsorbing macromolecules, opsonization of the intact complex by natural macromolecules coating over the targeting capacity of the complex, or simply that the nanoparticles were too big to leave the systemic circulation. In an attempt to ensure that the antibody assumed and maintained the correct surface orientation, Couvreur and Aubry[51] first conjugated protein A to the surface and then conjugated the antibody. When tested against H-29 colorectal carcinoma, they found that tumor uptake increased threefold compared to unconjugated nanoparticles. Different nanoparticles present different surface groups and therefore the conjugation procedure for both spacer and antibody has to be individually developed. Sezaki and Hashida[52] have recently reviewed conjugation procedures.

The final way to actively enhance targeting involves utilizing the inherent passive targeting ability of nontarget cells. Before the drug-containing nanoparticles are administered, a prior dose of blank nanoparticles, or similar nanoparticles containing other drugs, such as local anesthetics, could be administered to temporarily block off the actively phagocytic nontarget cells. In this way the less actively, but still somewhat, phagocytic target cells are able to interact with the desired drug delivery system. On the other hand, streptozotocin, used to make rats diabetic, also appears to cause hyperphagocytosis of the RES.[53] No reports have been made on using all these active targeting procedures together in an attempt to obtain better controlled delivery of any type of nanoparticles. When this is done, the true potential of the nanoparticle concept can be developed into commercially successful products.

The variety of acrylate macromolecules that can be used to form nanoparticles has been discussed in this section. Also, the various targeting procedures devised to better control delivery and release of the payload of drug have been outlined. In the remaining sections, the first of the questions to be answered in fabricating a controlled release drug delivery system will be discussed. Only when a different procedure to enhance the association of the delivery system and its site of action has been developed for these different types of nanoparticles will it be answered. The general procedures just described for acrylate nanoparticles apply equally to the other types of nanoparticles.

IV. PROTEINACEOUS NANOPARTICLES

From the large number of proteins which could be used as the base macromolecule from which to make nanoparticles, the system limitations discussed in Section II almost completely eliminate all possible proteins. The immune system of the body has the prime function of recognizing and rejecting proteins which are foreign to the orderly functioning of the body. Human serum albumin and gelatin seem to be the only two proteins which have any real hope of not causing a severe immunological response or reaction.

A. Albumin Nanoparticles

Human albumin products made for therapeutic and diagnostic use following parenteral administration are based on albumins[54] that have been heated at 60°C for 10 hr to inactivate hepatitis and possibly other viruses. To minimize changes in the protein during this heating procedure, a combination of two approved stabilizers is usually added.[55] It is normally assumed that these are present in such small proportions that they have no effect on albumin nanoparticle production. No detailed study appears to have been done. Albumin microspheres and nanoparticles are usually made in one of three basic ways, each of which aims to produce a monodisperse hardened or denatured particulate albumin system.

1. Diagnostic Nanoparticles

For drug-free particles intended for subsequent surface conjugation with a diagnostic material such as ^{99m}Tc, a simple thermal denaturation of a protein aerosol in gas[56] or a two-step procedure of making microspheres in an air-driven spinning top and then thermally denaturing the microspheres in olive oil is used.[57] Millar et al.[57] show that standard stannous chloride reduction to produce the surface Tc label can also produce an impurity, Tc-Sn colloid. This problem is not addressed by any groups which Tc label their drug-containing nanoparticles and microspheres to prove the targeting ability of the drug delivery system. Pittard et al.[58] showed that very small amounts of Tc were excreted in human milk and that breast-feeding could safely resume 24 hr after administration of a Tc-conjugated macroaggregated albumin used for a lung scan. This controlled nonrelease of ''drug'' removes one nontarget fluid from consideration of in vivo behavior of these systems.

2. Emulsion Manufacture

The second most common way of making albumin particles involves either thermal denaturation or chemical cross-linking of albumin in a water-in-oil (w/o) emulsion. The thermal denaturation method has been the most popular since it was first developed by Rhodes et al.[59] in 1968. Initially 5- to 65-μm microspheres were produced, but nanoparticles with a diameter of approximately 1 μm or less were soon made by Zolle et al.[60] and Scheffel et al.[61] Despite many research groups and industrial manufacturers using the Scheffel et al. method (or with minor variations), it was not until 1984 that the first systematic, quantitative examination of the variables in the procedures for preparing albumin nanoparticles was published by Gallo et al.[62]

As in most reports on making albumin nanoparticles (commonly called microspheres, macroaggregates, or microaggregates), Gallo et al.[62] initially dissolved 125 mg of bovine serum albumin in 0.5 mℓ of water. It is considered unlikely that the use of bovine rather than human serum albumin will change the general manufacturing guidelines arising from the work. This aqueous solution was then emulsified into 30 mℓ of oil at 4°C by ultrasonication. The emulsion could be made by stirring, and guidelines will be given when chemical cross-linking procedures are discussed. The emulsion was then added dropwise to 100 mℓ of stirred oil at 125° in a 85-mm-diameter round-bottom flask. After denaturation had occurred for a defined time, the system was cooled to room temperature and could be cleaned up by ether washes, centrifuging, and collection on polycarbonate filters. The mean nanoparticle size ranged from 440 to 820 nm, with a range in standard deviation of 280 to 820 nm. Hence the procedure can withstand considerable variation without causing dramatic change in particle size. The nanoparticles produced are too big to cross fenestrated membranes and too small to be useful embolic delivery systems. However, a number of interesting guidelines were proposed. An increase in albumin concentration in the aqueous phase tends to produce a slightly (10%) smaller nanoparticle, and it does produce proportionally more nanoparticles per batch. More powerful, but not longer, ultrasonication produced smaller (20%) nanoparticles. At 2 mℓ of aqueous phase in 30 mℓ of oil, and to a certain extent at 1 mℓ in 30 mℓ, the hardened nanoparticle pellet upon centrifugation also contained large proteinaceous rocks with some entrapped oil. This then provides an upper limit to the number of particles that can be produced in a batch. If the hot oil had a temperature in the range 105 to 145°C, the nanoparticles had a smaller mean as the temperature rose. At 165°C it was suggested that the albumin instantaneously hardened at the w/o interface, leading to a larger, more porous particle. Cottonseed oil produced smaller nanoparticles than maize oil or light paraffin oil. An attempt to use 1-octanol as the oil, so that the resultant nanoparticle surface might be more hydrophilic, led to very large and variably sized particles.

Chemical cross-linking by 2,3-butadione or 1,5-glutaraldehyde has been used by Tomlinson et al.[63] as they made albumin nanoparticles from a w/o emulsion formed by stirring. A flat-bottomed glass beaker (diameter of 60 mm and height of 110 mm) equipped with four 4-mm-deep baffles positioned against the sides was used: 125 mℓ of olive oil was stirred by a 4-bladed axial-flow impeller with a 3-mm gap between the baffles and the impeller blades; 0.5 mℓ of an aqueous solution of albumin (25%) and drug in isotonic buffer at pH 7 are added dropwise to the oil. They found that as the impeller speed increased, the mean diameter of the microspheres decreased. If nanoparticles were required, a much more viscous oil has to be used. If the cell has no baffles, then the well-defined size maximum and Gaussian size distribution characteristics are lost, probably because the system can no longer be regarded as being in the transitional stage between laminar flow and fully developed turbulence. Many commercial products seem to be made in an unbaffled container, up to a few liters in size, as there can be relatively poor control of the particle size. Sperber and Johansson[64] have developed a sucrose density gradient system to obtain more uniform sizes. These microspheres and nanoparticles can be cleaned up with ether washes, centrifugation, and collection on a filter. Alternatively, and more usefully as far as making a controlled release product is concerned, the particles can be stabilized by chemical cross-linking. This can be done with small amounts (about 10% of the total volume) of 2,3-butadione and under gentle stirring conditions.

Alternatively, 0.1 mℓ of 5 to 25% aqueous glutaraldehyde solution can be stirred in under the same vigorous conditions used to make the emulsion or added as a 12.5% solution in toluene if a chloroform-toluene mixture is used instead of the oil.[65] No attempts are made to remove the excess cross-linking agent. Consequently,[33] the microspheres and nanoparticles made by this method, especially those with more than 4% glutaraldehyde,[66] are subject to continued after-hardening, variable biodegradation, variable drug release, unexpected in vitro cell toxicity, and in vivo adverse reactions. Excess residual aldehydic cross-linking agents can be removed by the addition of sodium sulfite or sodium metabisulfite.[67]

Albumin nanoparticles can be characterized by sedimentation field-flow fractionation[68] as well as conventional SEM. Albumin microspheres stored in plasma have a low (-7 to -9 mV) zeta potential.[63]

Tomlinson et al.[63] showed that their freeze-dried human serum albumin microspheres swell in an aqueous environment. If glutaraldehyde is used as the cross-linker, provided it had an initial aqueous concentration of more than 1%, then the extent of swelling is only dependent on the amount of drug in the microsphere. Using the example of a payload of 25% of the water-soluble sodium cromoglycate, the diameter increased 75% when placed in saline for 24 hr, compared to drug-free microspheres which only swelled 40%. This swelling has been well known for some time. Zolle et al.[69] found that microspheres made at 118°C showed a 50% average increase in diameter after 5 min, whereas microspheres prepared at 146°C only showed a 20% increase in 5 min, but continued to swell a further 13% in the next hour. This swelling will be a major problem in passive targeting by size characteristics.[63] It will also mean that the cross-linked albumin matrix is becoming more porous and this will influence the release characteristics of the payload of drug. It is not known if the much smaller nanoparticles will have the same swelling problem. There have been many reports and some detailed studies on the in vitro and in vivo release of drugs from albumin microspheres and nanoparticles.

It has been obvious since the very early work of Kramer[70] that drug can be released in at least three ways from albumin nanoparticles. Using 200- to 1200-nm particles containing a 0.4 and 0.8% payload of mercaptopurine and daunomycin, respectively, Kramer found that 40% of that payload could be exchanged from surface sites. However, 16% of the payload could be removed from the nanoparticles into the buffer in 2 hr, the remaining 84% releasing into the buffer solution very slowly. This biexponential in vitro release has been studied by Tomlinson et al.[63] in terms of the manufacturing variables. They attribute the rapid burst effect to release from the surface sites and to the second slower stages being controlled by the swelling of the spheres, solution of the drug, and diffusion of drug out of the resultant porous matrix. They believe that 15- to 25-μm microspheres made by their method may still contain a useful, slowly released payload of drug for effective therapy by embolic targeting. However, drugs will be released too quickly from their type of albumin nanoparticle to be clinically useful. Allowing the 1% glutaraldehyde to cross-link for increasing time reduces the burst effect and enables an extended release of sodium cromoglycate.

Lee et al.[71] made 100- to 200-μm bovine serum albumin microspheres cross-linked with 1% glutaraldehyde and with a 5 to 30% payload of progesterone. In vitro release studies showed a drop in drug release rate as the concentration of glutaraldehyde was increased to 3%. However, only the 1% glutaraldehyde cross-linked beads were susceptible to chymotrypsin digestion and so are the only candidates for in vivo testing. Because no time between manufacture and administration is mentioned, any after-hardening effects cannot be identified. Following intramuscular or subcutaneous injection of similar nanoparticles based on rabbit serum albumin into rabbits, there was a 2-day period of high drug levels as the burst period occurred. After that, serum drug levels were maintained at about 1 ng/mℓ for at least 20 days by a near zero-order rate of release.

However, Morimoto et al.[72,73] investigated in vitro release in the presence of protease or lysosomal fraction as well as in tumor-bearing animals from 1.4-μm albumin nanoparticles containing a payload of 3.3% 5-fluorouracil or 15% doxorubicin. The fluorouracil product burst released about 15% of its payload within 2 hr into buffer and then slowly released the remainder over many weeks. However, in the presence of protease, about 70 to 80% of the payload was rapidly released and the remaining much smaller amount was released more slowly. Hence attempting to predict and optimize release rates at the targeted site solely on the results for nonenzymic-containing in vitro systems is not likely to be useful. In vivo, essentially all free drug was removed from the tumors within 1 day, whereas more than 80% of the drug in the nanoparticles remained in the target cells for more than 1 week. By radiolabeling the albumin, Morimoto et al. showed that the rate of drug removal was only very slightly faster than the in vivo degradation of the nanoparticle. The drug in nanoparticle-treated animals survived significantly longer than control or free-drug treated animals.

3. Nanoparticles by Desolvation Techniques

Many microcapsule products are based on the competitive desolvation of a macromolecule by smaller solute molecules. By initially molecularly associating drugs with up to 500 mg of a macromolecule such as albumin by ''protein-binding'', and then desolving the complex to just the isotropic side of the coacervation phase boundary, the system can be visualized as being almost discrete rolled-up balls of drug-containing protein. These embryonic nanoparticles can be fixed in this configuration by adding formaldehyde or, preferably, glutaraldehyde. Detailed description of the different aspects of the manufacture of this type of nanoparticle is available elsewhere.[1,33,66,67] The extent of desolvation and approach to the phase boundary is monitored by nephelometry. As the boundary is approached, there is a sharp rise in the intensity of the scattered light just before the system becomes visibly turbid and coacervation has occurred. Resolvation of the macromolecules in this still essentially aqueous system can be done by the addition of solvents such as isopropanol or water, which cause a corresponding rapid drop in the scattered light. An excess of glutaraldehyde is added when the nephelometry reading indicates that the macromolecule is just outside this pre-coacervation rapidly changing light-scattering state. After hardening for a predetermined time under vigorous stirring conditions, the excess cross-linking agent is removed by addition reaction with sodium metabisulfite.[67] The nanoparticles are separated from free drug and low molecular weight impurities by size fraction chromatography on Sephadex® G50m.

Albumin nanoparticles with a diameter in the range of 150 to 1000 nm are readily made, although absolute control and reproducibility of particle size is not as good as for other types of nanoparticles. Consequently when it is important that a particle size fraction of nanoparticles be used, further size exclusion chromatography is done. It is proving difficult to reproducibly make proteinaceous nanoparticles by this method with a size less than 100 nm as would be required to cross from the general circulation and target, for example, tumor cells.[50]

Even when examined at a magnification of 30,000 × by SEM, there is no evidence that these nanoparticles are porous. Hence for the molecularly dispersed drug to be released in vivo, the nanoparticles must be enzymatically degraded. El-Samaligy and Rohdewald[74] have shown that gelatin nanoparticles can be readily degraded by trypsin, collaginase, and trypsin/collaginase mixtures. However, it has been shown[75] that refluxing water-soluble flukicide-containing albumin nanoparticles in 0.01 M NaOH or HCl for 1 day did not cause the release of more than 2% of a water-soluble flukicide. As these submicron drug delivery systems are made to target particular cellular sites, control of drug release depends upon nanoparticle biodegradation rather than in vitro tests.

Marty[76] found that the inclusion of a surface-active agent in the starting system improved the dispersibility of the final product, even though most of the surfactant is separated from the nanoparticles in the purification step. Polysorbate 80 was originally used, but this was found to be unsuitable since its cloud point was lowered upon the addition of the desolvating agent to below the working temperature of 35°C. The resulting cloudiness interfered with the nephelometer monitoring of the desolvation of the albumin. Polysorbate 20 was found not to interfere with the monitoring and is routinely used by groups making this type of nanoparticle.[42,74,78] The surfactant also assisted in transforming water-insoluble drugs into a solubilized form from which they could molecularly interact with the protein to ensure a workable payload.[1,43,75]

Payloads achieved have varied[79] from 0.2% melphalan in human serum albumin to almost 16% potassium triamcinolone acetonide 21-phosphate in rabbit serum albumin. It is difficult to conceive of payloads of more than about 25% w/w without aggregates of drug in the nanoparticle or pores rapidly developing as the result of release of drug molecules close together. Scale-up for this type of nanoparticle would require adequate control of the stirring rate during the desolvation and hardening steps. Although the glutaraldehyde addition would assist in large-scale sterile manufacture,[42] adequate safety assurance would have to be demonstrated, particularly with regard to viral contamination.

4. Adverse Reactions

Polysorbate 20 and 80 have been reported[80-82] to release endogenous stores of histamine, increase capillary permeability, and result in an anaphylactic-like syndrome. These reactions which could occur when Polysorbate 80 is used as an adjuvant for the microspheres are related to Polysorbate 80 and not the microspheres.[82]

The greatest concern with controlled release drug delivery systems based on albumin is the induction of immunogenicity as a result of the manufacturing procedure. Onica et al.[83] conclude that glutaraldehyde treatment of rabbit albumin results in the monomeric albumin being antigenically identical to albumin altered by "aging" during storage. There was, however, a new haptenic determinant present on the polymerized albumin. On the other hand, Strambachova-McBride et al.[84] believe that rather than a new antigenic determinant being exposed on heat-aggregated bovine serum albumin, there is an enhanced capacity of polymerized bovine serum albumin to establish polyvalent binding to an antibody of low intrinsic affinity.

Keeling[85] has recently reviewed adverse reactions to radiopharmaceuticals administered in the U.K. and U.S. and an incidence figure of about 1 in 6000 can be derived for albumin-based products. However, Rhodes and Cordova[86] consider that microspheres are ten times more likely to cause a response than macroaggregated albumin. Only future clinical experience will confirm if extrapolation of this trend to albumin nanoparticles is likely to cause an even bigger incidence and if the use of these systems to target drugs modulates the incidence rate.

5. Controlled Release by Targeting

Perhaps the most imaginative approach to targeting nanoparticles was developed by Widder et al.[87] Using the phase separation emulsion polymerization method, 10- to 20-nm particles of magnetite are incorporated in the aqueous albumin and drug-containing solution and after sonication the nanoparticles are hardened either by heat denaturation or chemical cross-linking. The temperature of hardening is important[1] in that for doxorubicin-containing nanoparticles, stabilization at 115°C produced no drug degradation, whereas only 78% of the incorporated material remained biologically active after stabilization of the nanoparticle at 135°C. In vitro, 20% of the residual active drug in the 135°C product was dumped in 1 hr, but then the remainder was zero-order released at about 6%/hr. As the nanoparticles degraded only about 3% in 48 hr,[88] the drug release appears to correspond with the rate of nanoparticle hydration. As the manufacturing procedure could alter this, some control on the release rate is possible. The release rate could also be dependent on the diffusion characteristics of the incorporated drug.

Retention of the magnetically responsive nanoparticles with their payload of drug in the microvasculature can be achieved[89] by taking advantage of the difference in the linear flow velocity of blood in a large artery (15 to 30 cm/sec) compared to that in capillaries (0.05 cm/sec). A much smaller magnetic field is required to retain the nanoparticles in the capillaries than in the arteries. This specific site entrapment enables drug to be released into capillaries and venules which have a high drug diffusibility. There can be no possibility of nanoparticles crossing out of the circulation into the extravascular space and perhaps even being taken up by target cells. However, this local perfusion technique has enabled primary tumors with well-developed blood supplies to be treated. It cannot be used to treat metastatic deposits.

B. Gelatin Nanoparticles

There are three major differences in albumin and gelatin which influence the usefulness of controlled release nanoparticles based on these proteins. First, although reliable data on drug-gelatin interactions are difficult to find, it is generally accepted that protein binding of drugs to gelatin is weaker than that to albumin. Second, gelatin is gelled by cooling, whereas albumin is "gelled" or made into particles by heating. Hence the preparation of gelatin-based particulate systems is less likely to lead to drug degradation caused by elevated temperatures. Third, gelatin nanoparticles, although causing some immunological response,[33] appear to be less antigenic than albumin nanoparticles. In the immunologically sensitive procedure of intra-articular injection, Kennedy[43] found that drug-laden and drug-free gelatin nanoparticles caused essentially no joint inflammation or change in joint diameter upon repeated injection in the knee joints of rabbits. As expected, similar bovine serum albumin nanoparticles did cause marked responses on second and subsequent injection.

Gelatin nanoparticles can be made by either the emulsion[77] or desolvation[67] methods followed by chemical cross-linking as described for albumin nanoparticles. In addition, gelatin microspheres can be made by grinding up gelatin sponge.[33] Harrington and Barry[78] found that hardening desolvated gelatin with glutaraldehyde for more than 4 min caused an increase in the hydrophobic character of the surface. This very interesting in vitro result could mean that in vivo there will be a variety of opsonization processes on the surface of the nanoparticle which could lead to uptake by a variety of cells. It is possible that a similar effect could be found for albumin nanoparticles.

Chen et al.[90] were able to target 0.2-mm magnetic gelatin beads containing 5-fluorouracil to a target site in the human esophagus by application of an external magnet. The concentration of drug in esophageal mucosa was over 30 times that which could be achieved by intravenous infusion. Similarly, the effect of different surfactants[30,50] on the size and targeting ability of gelatin nanoparticles needs to be assessed.

Control of drug release follows the same guidelines as for albumin nanoparticles. Yoshioka et al.[77] made 280-nm nanoparticles and 15-μm microspheres by the emulsion technique followed by chemically hardening the gelatin with formaldehyde. Payloads of around 2% of mitomycin-C as both free drug and in the form of its dextran conjugate were achieved.

More than 50% of the payload of the drug in the nanoparticles was released in vitro at 37°C within 10 min. This implies a very porous product. On the other hand, it took about 2 hr for the release from the 15-μm microspheres. Assuming a similar porosity in the two products, these release rates are similar if allowance is made for relative surface area and volume of the two sizes of particles. The release of the dextran conjugate from either size particle followed monoexponential kinetics with a half-life of about 30 hr which is similar to that for hydrolytic cleavage of mitomycin-C from dextran.

Preliminary experiments[102] show that intravenous preadministration of unlabeled drug-free and lignocaine-containing gelatin nanoparticles could slow down the rate of liver accumulation of Tc-labeled gelatin nanoparticles administered intravenously into rats. As would be expected, the rate of liver accumulation following intramuscular and intraperitoneal administration of Tc-labeled 400-nm gelatin nanoparticles is much slower than that following intravenous administration.[91] Fluorescein isothiocyanate can be conjugated onto the surface of gelatin nanoparticles.[41] A variety of standard test tumor cell lines internalize this conjugate system, e.g., EMT-6, WEHI-3, and SP-1[41] and B16 and the R111 viral mouse mammary tumor.[92] Other phagocytic cells such as peritoneal polymorphonuclear cells[103] internalize the nanoparticles and surface washing of the cells produces very little fluorescence. There seems to be no reason why antibodies could not be conjugated to the same surface sites on the gelatin nanoparticles in an attempt to increase the targeting capacity of the system. In summary, gelatin seems to be an ideal base macromolecule to use to fabricate a targetable controlled release drug delivery system.

C. Insulin Nanoparticles

Oppenheim et al.[93] were able to make 200-nm nanoparticles by a modification of the desolvation method using insulin as the base macromolecule. Rather than using competing solutes to desolvate the macromolecule, the pH was adjusted so that the insulin was at the minimum in its pH/solubility profile. By carefully controlling the total concentration of the protein, nanoparticles were made. The nanoparticles had a potency of approximately 5 IU/mg when injected intravenously into rats. When administered orally at approximately 50 mg/100 g body weight, the blood glucose of some normal and diabetic rats was reduced to about 20% of the starting levels 3 hr after administration. Regrettably the high doses of nanoparticles needed precluded the development of a commercially viable product.

V. OTHER SYSTEMS

The many other particulate controlled released drug delivery systems can be readily divided by the biodegradation criteria. Most of the biodegradable particulate systems mentioned in this section could possibly be made as nanoparticles. Two examples of non- or poorly biodegraded systems will be given, but until a more realistic approach is taken to biodegradation or toxicity, none of the many such systems are likely to achieve commercial viability.

A. Other Biodegradable Particulate Systems

Pharmacia has been putting a lot of effort into their embolic biodegradable starch microspheres, Spherex®. Typical reviews of these 15-μm products have been given recently by Russell[94] and Lindberg et al.[95] The degradation half-life by systemic amylase can be varied from 10 to 100 min by changing the particle size and the relative degree of cross-linking. Again the control on size is achieved by control of stirring speed, stirring method, and vessel geometry[94] in the standard way for bead polymerization.

Schröder and Ståhl[96] entrapped antigens such as ovalbumin, bee venom, and schistosoma antigens in 1-μm carbohydrate spheres. A w/o emulsification procedure was used and the particle size generated by probe sonication. Antibody responses were seen in mice for up to 15 weeks. Polylactic acid has been used as a biodegradable implant[97] and particulate system[98] for the controlled release of drugs and hormones. Microspheres can be formed by first forming a water-in-organic phase emulsion in a Waring® blender and then evaporating the solvent. Although the emulsification is likely to cause particle size variation, the particle-particle aggregation that must occur during the evaporation of the solvent is likely to present major problems. A useful extension of the polylactic acid systems would be to make nanoparticles from this polymer.

The final group of potentially biodegradable nanoparticles could be seen as solid liposomes. Fendler and Tundo[99] indicate that the experimental development of polymerized lipid and surfactant nanoparticles is only a question of time and ingenuity. The Squibb proliposome concept is an example of recent ingenuity.

B. Other Nonbiodegradable and Possibly Toxic Systems

Doxorubicin can be coupled to 450-nm polyglutaraldehyde nanoparticles and still remain cytotoxic,[100] probably by interaction with target membranes. Ege[101] has been using Tc-labeled 10-nm antimony sulfur colloid to image lymph nodes. Potentially drugs could also be conjugated to the colloid and targeted to nodes containing metastases.

VI. SUMMARY

A rich variety of base molecules has been used to make solid submicron particles which try to target diseased cells, releasing their payload of drug at the correct site of action in a controlled manner. The manufacturing conditions needed to control the drug release once the target site has been reached, and only then, are in general well understood. The designers of such drug delivery systems are still attempting to answer the second question relating to the way the system interacts with the body before the system arrives at the desired release site. Until both questions are adequately answered, nanoparticles as a targeted controlled release drug delivery system will have to remain in the laboratory and experimental clinic and await their proper industrial and commercial exploitation in the preferably not-too-distant future.

ACKNOWLEDGMENTS

The financial support from the National Health and Medical Research Council (830092) and the Wool Research Trust Fund of the Australian Wool Corporation (M/2/1403) which assisted in the preparation of this chapter is gratefully acknowledged.

REFERENCES

1. **Oppenheim, R. C.,** Solid colloidal drug delivery systems: nanoparticles, *Int. J. Pharm.*, 8, 217, 1981.
2. **Widder, K. J., Senyei, A. E., and Ranney, D. F.,** Magnetically responsive microspheres and other carriers for the biophysical targeting of antitumour agents, *Adv. Pharmacol. Chemother.*, 16, 213, 1979.
3. **Tomlinson, E.,** Microsphere delivery systems for drug targeting and controlled release, *Int. J. Pharm. Technol. Prod. Manuf.*, 4, 49, 1983.
4. **Kreuter, J.,** Evaluation of nanoparticles as drug-delivery systems. I. Preparation methods, *Pharm. Acta Helv.*, 58, 196, 1983.
5. **Illum, L. and Davis, S. S.,** *Polymers in Controlled Drug Delivery*, Butterworths, Guilford, 1987.
6. **Davis, S. S., Illum, L., McVie, J. G., and Tomlinson, E., Eds.,** *Microspheres and Drug Therapy: Pharmaceutical, Immunological and Medical Aspects*, Elsevier, Amsterdam, 1984.
7. **Illum, L.,** In search of polysomes, *Pharm. Int.*, 5, 185, 1984.

8. **Kreuter, J. and Hartmann, H. R.**, Comparative study of the cytostatic effects and tissue distribution of 5-fluorouracil in a free form and bound to polybutylcyanoacrylate nanoparticles in Sarcoma 180-bearing mice, *Oncology*, 40, 363, 1983.

9. **Kreuter, J.**, Physicochemical characterization of polyacrylic nanoparticles, *Int. J. Pharm.*, 14, 43, 1983.

10. **Clark, D. S., Bailey, J. E., Yen, R., and Rembaum, A.**, Enzyme immobilization on grafted polymeric microspheres, *Enzyme Microb. Technol.*, 6, 317, 1984.

11. **Lindmark, R., Larsson, E., Nilsson, K., and Sjöquist, J.**, Immobilization of proteins by entrapment in polyacrylamide microbeads, *J. Immunol. Methods*, 49, 159, 1982.

12. **Ekman, B. and Sjöholm, I.**, Improved stability of proteins immobilized in microparticles prepared by a modified emulsion polymerization technique, *J. Pharm. Sci.*, 67, 693, 1978.

13. **Ekman, B., Lofter, C., and Sjöholm, I.**, Incorporation of macromolecules in microparticles: preparation and characteristics, *Biochemistry*, 15, 5115, 1976.

14. **Edman, P., Ekman, B., and Sjöholm, I.**, Immobilization of proteins in microspheres of biodegradable polyacryldextran, *J. Pharm. Sci.*, 69, 838, 1980.

15. **Sjöholm, I. and Edman, P.**, The use of biocompatible microparticles as carriers of enzymes and drugs in vivo, in *Microspheres and Drug Therapy: Pharmaceutical, Immunological and Medical Aspects*, Davis, S. S., Illum, L., McVie, J. G., and Tomlinson, E., Eds., Elsevier, Amsterdam, 1984, sect. 3, chap. 6.

16. **Artusson, P., Edman, P., Laakso, T., and Sjöholm, I.**, Characterization of polyacryl starch microparticles as carriers for proteins and drugs, *J. Pharm. Sci.*, 73, 1507, 1984.

17. **El-Samaligy, M. and Rohdewald, P.**, Polyacrylamide microbeads, a sustained release drug delivery system, *Int. J. Pharm.*, 13, 23, 1983.

18. **Edman, P. and Sjöholm, I.**, Acrylic microspheres in vivo. V. Immunological properties of immobilized asparaginase in microparticles, *J. Pharm. Sci.*, 71, 576, 1982.

19. **Edman, P., Sjöholm I., and Brunk, U.**, Acrylic microspheres in vivo. VII. Morphological studies on mice and cultured macrophages, *J. Pharm. Sci.*, 72, 658, 1983.

20. **Birrenbach, G. and Speiser, P. P.**, Polymerized micelles and their use as adjuvants in immunology, *J. Pharm. Sci.*, 65, 1763, 1976.

21. **Bentele, V., Berg, U. E., and Kreuter, J.**, Molecular weights of poly(methylmethacrylate) nanoparticles, *Int. J. Pharm.*, 13, 109, 1983.

22. **Candau, F., Leong, Y. S., Pouyet, G., and Candau, S.**, Inverse microemulsion polymerization of acrylamide: characterization of the water-in-oil microemulsions and the final microlatexes, *J. Colloid Interface Sci.*, 101, 167, 1984.

23. **Marty, J. J. and Oppenheim, R. C.**, Colloidal systems for drug delivery, *Aust. J. Pharm. Sci.*, 6, 65, 1977.

24. **Douglas, S. J., Illum, L., Davis, S. S., and Kreuter, J.**, Particle size and size distribution of poly(butyl-2-cyanoacrylate) nanoparticles. I. Influence of physicochemical factors, *J. Colloid Interface Sci.*, 101, 149, 1984.

25. **El-Egakey, M. A. and Speiser, P.**, Drug loading studies on ultrafine solid carriers by sorption procedures, *Pharm. Acta Helv.*, 57, 236, 1982.

26. **Kante, B., Couvreur, P., Lenaerts, V., Guiot, P., Roland, M., Baudhuin, P., and Speiser, P.**, Tissue distribution of [3H] actinomycin D adsorbed on polybutylcyanoacrylate nanoparticles, *Int. J. Pharm.*, 7, 45, 1980.

27. **Illum, L., Jones, P. D. E., Baldwin, R. W., and Davis, S. S.**, Tissue distribution of poly(hexyl-2-cyanoacrylate) nanoparticles coated with monoclonal antibodies in mice bearing human tumor xenografts, *J. Pharmacol. Exp. Ther.*, 230, 733, 1984.

28. **Illum, L., Jones, P. D. E., Kreuter, J., Baldwin, R. W., and Davis, S. S.**, Adsorption of monoclonal antibodies to polyhexylcyanoacrylate nanoparticles and subsequent immunospecific binding to tumour cells in vitro, *Int. J. Pharm.*, 17, 65, 1983.

29. **Goodwin, J. W., Ottewill, R. H., Petton, R., Vianello, G., and Yates, D. E.**, Control of particle size in the formation of polymer latexes, *Br. Polym. J.*, 10, 173, 1978.

30. **Douglas, S. J., Illum, L., and Davis, S. S.**, Particle size and size distribution of poly(butyl-2-cyanoacrylate) nanoparticles. II. Influence of stabilizer, *J. Colloid Interface Sci.*, 103, 154, 1985.

31. **Pani, K. C., Gladieux, G., Brandes, G., Kulkarni, R. K., and Leonard, F.**, The degradation of n-butyl alpha-cyanoacrylate tissue adhesive. II, *Surgery*, 63, 481, 1968.

32. **Kante, B., Couvreur, P., Dubois-Krack, G., De Meester, C., Guiot, P., Roland, M., Mercier, M., and Speiser, P.**, Toxicity of polyalkylcyanoacrylate nanoparticles. I. Free nanoparticles, *J. Pharm. Sci.*, 71, 786, 1982.

33. **Oppenheim, R. C.**, Gelatin microspheres and nanoparticles as drug carrier systems, in *Polymers in Controlled Drug Delivery*, Illum, L. and Davis, S. S., Eds., Butterworths, Guilford, 1987, chap. 6.

34. **Couvreur, P., Lenaerts, V., Leyh, D., Guiot, P., and Roland, M.,** Design of biodegradable polyalkyl-cyanoacrylate nanoparticles as a drug carrier, in *Microspheres and Drug Therapy: Pharmaceutical, Immunological and Medical Aspects,* Davis, S. S., Illum, L., McVie, J. G., and Tomlinson, E., Eds., Elsevier, Amsterdam, 1984, sect. 2, chap. 3.
35. **Brasseur, F., Couvreur, P., Kante, B., Deckers-Passau, L., Roland, M., Deckers, C., and Speiser, P.,** Actinomycin D adsorbed on polymethylcyanoacrylate nanoparticles: increased efficiency against an experimental tumour, *Eur. J. Cancer,* 16, 1441, 1980.
36. **Couvreur, P.,** Mise au Point d'un Nouveau Vecteur de Medicament, l'Agrégation de l'Enseignement Supérieur thesis, Université Catholique de Louvain, Bruxelles, 1983.
37. **Kreuter, J., Mills, S. N., Davis, S. S., and Wilson, C. G.,** Polybutylcyanoacrylate nanoparticles for the delivery of [^{75}Se] norcholestenol, *Int. J. Pharm.,* 16, 105, 1983.
38. **Couvreur, P., Kante, B., Lenaerts, V., Seailteur, V., Roland, M., and Speiser, P.,** Tissue distribution of antitumour drugs associated with polyalkylcyanoacrylate nanoparticles, *J. Pharm. Sci.,* 69, 199, 1980.
39. **Debrun, G., Vinuela, F., Fox, A., and Drake, C. G.,** Embolization of cerebral arteriovenous malformations with bucrylate, *J. Neurosurg.,* 56, 615, 1982.
40. **Ramchandani, P., Goldenberg, N. J., Soulen, R. L., and White, R. I.,** Isobutyl-2-cyanoacrylate embolization of a hepatoportal fistula, *Am. J. Radiol.,* 140, 137, 1983.
41. **Lenaerts, V., Nagelkerke, J. F., van Berkel, T. J. C., Couvreur, P., Grislain, L., Roland, M., and Speiser, P.,** In vivo uptake of polyisobutylcyanoacrylate nanoparticles by rat liver Kupffer, endothelial and parenchymal cells, *J. Pharm. Sci.,* 73, 980, 1984.
42. **Oppenheim, R. C. and Stewart, N. F.,** The manufacture and tumour cell uptake of nanoparticles labelled with fluorescein isothiocyanate, *Drug Dev. Ind. Pharm.,* 5, 563, 1979.
43. **Kennedy, K. T.,** Preparation and Testing of Nanoparticles Containing Triamcinolone Acetonide, M.Pharm. thesis, Victorian College of Pharmacy, Ltd., Parkville, Australia, 1983.
44. **Couvreur, P., Lenaerts, V., Kante, B., Roland, M., and Speiser, P.,** Oral and parenteral administration of insulin associated with hydrolysable nanoparticles, *Acta Pharm. Technol.,* 26, 220, 1980.
45. **Sandy, J. D., Sriratana, A., Brown, H. L. G., and Lowther, D. A.,** Evidence for polymorphonuclear-leucocyte-derived proteinases in arthritic cartilage, *Biochem. J.,* 193, 93, 1981.
46. **El-Samaligy, M. and Rohdewald, P.,** Triamcinolone diacetate nanoparticles, a sustained release drug delivery system suitable for parenteral administration, *Pharm. Acta Helv.,* 57, 201, 1982.
47. **Grislain, L., Couvreur, P., Lenaerts, V., Roland, M., Deprez-Decampeneere, D., and Speiser, P.,** Pharmacokinetics and distribution of a biodegradable drug-carrier, *Int. J. Pharm.,* 15, 335, 1983.
48. **Kreuter, J., Nefzger, M., Liehl, E., Czok, R., and Voges, R.,** Distribution and elimination of poly(methylmethacrylate) nanoparticles after subcutaneous administration to rats, *J. Pharm. Sci.,* 72, 1146, 1983.
49. **Ibrahim, A., Couvreur, P., Roland, M., and Speiser, P.,** New magnetic drug carrier, *J. Pharm. Pharmacol.,* 35, 59, 1983.
50. **Illum, L. and Davis, S. S.,** The organ uptake of intravenously administered colloidal particles can be altered using a non-ionic surfactant (Poloxamer 338), *FEBS Lett.,* 167, 79, 1984.
51. **Couvreur, P. and Aubry, J.,** Monoclonal antibodies for the targeting of drugs: application to nanoparticles, in *Topics in Pharmaceutical Sciences 1983,* Breimer, D. D. and Speiser, P., Eds., Elsevier, Amsterdam, 1983, 305.
52. **Sezaki, H. and Hashida, M.,** Macromolecule-drug conjugates in targeted cancer chemotherapy, *CRC Crit. Rev. Ther. Drug Carrier Syst.,* 1, 1, 1984.
53. **Cornell, R. P.,** Reticuloendothelial hyperphagocytosis occurs in streptozotocin-diabetic rats. Studies with colloidal carbon, albumin microaggregates and soluble fibrin monomers, *Diabetes,* 31, 110, 1982.
54. Code of U.S. Federal Regulations, Title 21, Parts 640.80 to 640.86 and 640.90 to 640.96, 1982.
55. **Yu, M. W. and Finlayson, J. S.,** Quantitative determination of the stabilizers octanoic acid and N-acetyl-DL-tryptophan in human albumin products, *J. Pharm. Sci.,* 73, 82, 1984.
56. **Przyborowski, M., Lachnik, E., Wiza, J., and Licińska, I.,** Preparation of HSA microspheres in a one-step thermal denaturation or protein aerosol carried out in a gas medium, *Eur. J. Nucl. Med.,* 7, 71, 1982.
57. **Millar, A. M., McMillan, L., Hannan, W. J., Emmett, P. C., and Aitken, R. J.,** The preparation of dry, monodisperse microspheres of [^{99m}Tc] albumin for lung ventilation imaging, *Int. J. Appl. Radiat. Isot.,* 33, 1423, 1982.
58. **Pittard, W. B., III, Merkatz, R., and Fletcher, B. D.,** Radioactive excretion in human milk following administration of technetium Tc-99m macroaggregated albumin, *Pediatrics,* 70, 231, 1982.
59. **Rhodes, B. A., Zolle, I., and Wagner, H. N., Jr.,** Properties and use of radioactive albumin microspheres, *Clin. Res.,* 16, 245, 1968.
60. **Zolle, I., Hosain, F., Rhodes, B. A., and Wagner, H. N., Jr.,** Human serum albumin microspheres for studies of the reticuloendothelial system, *J. Nucl. Med.,* 11, 379, 1970.
61. **Scheffel, U., Rhodes, B. A., Natarajan, T. K., and Wagner, H. N., Jr.,** Albumin microspheres for study of the reticuloendothelial system, *J. Nucl. Med.,* 13, 498, 1972.

62. **Gallo, J. M., Hung, C. T., and Perrier, D. G.,** Analysis of albumin microsphere preparation, *Int. J. Pharm.,* 22, 63, 1984.

63. **Tomlinson, E., Burger, J. J., Schoonderwoerd, E. M. A., and McVie, J. G.,** Human serum albumin microspheres for intraarterial drug targeting of cytostatic compounds. Pharmaceutical aspects and release characteristics, in *Microspheres and Drug Therapy: Pharmaceutical, Immunological and Medical Aspects,* Davis, S. S., Illum, L., McVie, J. G., and Tomlinson, E., Eds., Elsevier, Amsterdam, 1984, sect. 2, chap. 1.

64. **Sperber, G. O. and Johansson, A.,** A method to improve the size uniformity of microspheres, *Acta Physiol. Scand.,* 109, 111, 1980.

65. **Longo, W. E., Iwata, H., Lindheimer, T. A., and Goldberg, E. P.,** Preparation of hydrophilic albumin microspheres using polymeric dispersing agents, *J. Pharm. Sci.,* 71, 1323, 1982.

66. **Oppenheim, R. C.,** Nanoparticulate drug delivery systems based on gelatin and albumin, in *Polymeric Microparticles,* Guiot, P. and Couvreur, P., Eds., CRC Press, Boca Raton, Fla., 1985, chap. 2.

67. **Pharmaceutical Society of Victoria and Speiser, P.,** Injectable Compositions, U.S. Patent 4,107,288, 1978.

68. **Caldwell, K. D., Karaiskakis, G., Myers, M. N., and Giddings, J. C.,** Characterization of albumin microspheres by sedimentation field-flow fractionation, *J. Pharm. Sci.,* 70, 1350, 1981.

69. **Zolle, I., Rhodes, B. A., and Wagner, H. N., Jr.,** Preparation of metabolizable radioactive human serum albumin microspheres for studies of the circulation, *Int. J. Appl. Radiat. Isot.,* 21, 155, 1970.

70. **Kramer, P. A.,** Albumin microspheres as vehicles for achieving specificity in drug delivery, *J. Pharm. Sci.,* 63, 1646, 1974.

71. **Lee, T. K., Sokoloski, T. D., and Royer, G. P.,** Serum albumin beads: an injectable, biodegradable system for the sustained release of drugs, *Science,* 213, 233, 1981.

72. **Morimoto, Y., Akimoto, M., Sugibayashi, K., Nadai, T., and Kato, Y.,** Drug carrier property of albumin microspheres in chemotherapy. IV. Antitumor effect of single-shot or multiple-shot administration of microsphere-entrapped 5-fluorouracil on Ehrlich ascites or solid tumour in mice. *Chem. Pharm. Bull.,* 28, 3087, 1980.

73. **Morimoto, Y., Sugibayashi, K., and Kato, Y.,** Drug carrier property of albumin microspheres in chemotherapy. V. Antitumour effect of microsphere-entrapped adriamycin on liver metastasis of AH 7974 cells in rats, *Chem. Pharm. Bull.,* 29, 1433, 1981.

74. **El-Samaligy, M. S. and Rohdewald, P.,** Reconstituted collagen nanoparticles, a novel drug carrier delivery system, *J. Pharm. Pharmacol.,* 35, 537, 1983.

75. **Boag, C. C.,** Organ Directive Drug Delivery Systems, M.Pharm. thesis, Victoria Institute of Colleges, Melbourne, Australia, 1979.

76. **Marty, J. J.,** The Preparation, Purification, and Properties of Nanoparticles, D.Pharm. thesis, Pharmaceutical Society of Victoria, Parkville, Australia, 1977.

77. **Yoshioka, T., Hashida, M., Muranishi, S., and Sezaki, H.,** Specific delivery of mitomycin C to the liver, spleen, and lung: nano- and microspherical carriers of gelatin, *Int. J. Pharm.,* 8, 131, 1981.

78. **Harrington, R. and Barry, B. W.,** Hydrophobic interaction chromatography of bovine serum albumin nanoparticles, *J. Pharm. Pharmacol.,* 34, 79P, 1982.

79. **Oppenheim, R. C., Gipps, E. M., Forbes, J. F., and Whitehead, R. H.,** Development and testing of proteinaceous nanoparticles containing cytotoxics, in *Microspheres and Drug Therapy: Pharmaceutical, Immunological and Medical Aspects,* Davis, S. S., Ilum, L., McVie, J. G., and Tomlinson, E., Eds., Elsevier, Amsterdam, 1984, sect. 2, chap. 4.

80. **Burnell, R. H. and Maxwell, G. M.,** General and coronary hemodynamic effects of Tween 20, *Aust. J. Exp. Biol. Med. Sci.,* 52, 151, 1974.

81. **Marks, L. S. and Kolman, S. N.,** Tween 20 shock in dogs and related fibrinogen changes, *Am. J. Physiol.,* 220, 218, 1971.

82. **Millard, R. W., Bonig, H., and Vatner, S. F.,** Cardiovascular effects of radioactive microsphere suspension and Tween 80 solutions, *Am. J. Physiol.,* 232, H331, 1977.

83. **Onica, D., Dobre, M-A., and Lenkei, R.,** Immunogenicity of glutaraldehyde treated homologous monomeric albumin in rabbits, *Immunochemistry,* 15, 941, 1978.

84. **Strambachova-McBride, J., McBride, W. H., and Weir, D. M.,** Are "new" antigenic determinants exposed on aggregated bovine serum albumin?, *Clin. Exp. Immunol.,* 39, 233, 1980.

85. **Keeling, D. H.,** Adverse reactions to radiopharmaceuticals, in *Safety and Efficacy of Radiopharmaceuticals,* Kristensen, K. and Nørbygaard, E., Eds., Martinus Nijhoff, Boston, 1984, chap. 16.

86. **Rhodes, B. A. and Cordova, M. A.,** Adverse reactions to radiopharmaceuticals: incidence in 1978 and associated symptoms, *J. Nucl. Med.,* 21, 1107, 1980.

87. **Widder, K. J., Senyei, A. E., and Scarpelli, D. G.,** Magnetic microspheres: a model system for site-specific drug delivery in vivo, *Proc. Soc. Exp. Biol. Med.,* 158, 141, 1978.

88. **Widder, K. J., Flouret, G., and Senyei, A. E.,** Magnetic microspheres: synthesis of a novel parenteral drug delivery, *J. Pharm. Sci.,* 68, 79, 1979.

89. **Widder, K. J., Senyei, A. E., and Sears, B.,** Experimental methods in cancer therapeutics, *J. Pharm. Sci.,* 71, 379, 1982.
90. **Chen, Q., Sun, S-Y., Gu, X-Q., Li, Z-W., and Li, Z-X.,** Magnetic microspherical carrier — a new dosage form of fluorouracil as an anticancer agent, *Acta Pharmacol. Sin.,* 4, 273, 1983.
91. **Oppenheim, R. C.,** Nanoparticles, in *Drug Delivery Systems: Characteristics and Biomedical Applications,* Juliano, R. L., Ed., Oxford University Press, New York, 1980, chap. 5.
92. **Gipps, E. M.,** The Incorporation of Cytotoxics into Nanoparticles, M.Pharm. thesis, Victorian College of Pharmacy, Ltd., Parkville, Australia, 1983.
93. **Oppenheim, R. C., Stewart, N. F., Gordon, L., and Patel, H. M.,** The production and evaluation of orally administered insulin nanoparticles, *Drug Dev. Ind. Pharm.,* 8, 531, 1982.
94. **Russell, G. F. J.,** Starch microspheres as drug delivery systems, *Pharm. Int..,* 3, 260, 1983.
95. **Lindberg, B., Lote, K., and Teder, H.,** Biodegradable starch microspheres — a new medical tool, in *Microspheres and Drug Therapy: Pharmaceutical, Immunological and Medical Aspects,* Davis, S. S., Illum, L., McVie, J. G., and Tomlinson, E., Eds., Elsevier, Amsterdam, 1984, sect. 3, chap. 1.
96. **Schröder, U. and Ståhl, A,** Crystallized dextran nanospheres with entrapped antigen and their use as adjuvants, *J. Immunol. Methods,* 70, 127, 1984.
97. **Yolles, S., Leafe, T., Ward, L., and Boettner, F.,** Controlled release of biologically active drugs, *Bull. Parenter. Drug Assoc.,* 30, 306, 1976.
98. **Sanders, C. M., Kent, J. S., McRae, G. I., Vickery, B. H., Tice, T. R., and Lewis, D. H.,** Controlled release of a luteinizing hormone-releasing hormone analogue from poly(d,1-lactide-co-glycolide) microspheres, *J. Pharm. Sci.,* 73, 1294, 1984.
99. **Fendler, J. H. and Tundo, P.,** Polymerized surfactant aggregates: characterization and utilization, *Acc. Chem. Res.,* 17, 3, 1984.
100. **Tokes, Z. A., Rogers, K. E., and Rembaum, A.,** Synthesis of adriamycin-coupled polyglutaraldehyde microspheres and evaluation of their cytostatic activity, *Proc. Natl. Acad. Sci. U.S.A.,* 79, 2026, 1982.
101. **Ege, G. N.,** Internal mammary lymphoscintography: a rational adjunct to the staging and management of breast carcinoma, *Clin. Radiol.,* 29, 453, 1978.
102. **Oppenheim, R. C., Lichtenstein, M., Chang, J. M., and Chieng, L. K.,** unpublished results, 1984.
103. **Oppenheim, R. C. and Lowther, D. A.,** unpublished results, 1983.

Chapter 2

MULTIPLE EMULSIONS

Gary H. Cueman and Joel L. Zatz

TABLE OF CONTENTS

I. INTRODUCTION

An emulsion is a complex system made up of two or more immiscible fluid phases. Most emulsions contain only two phases, one of which is usually water. The other liquid is given the generic title "oil". With such a two-component system, it is possible to have oil droplets dispersed in water (described as an oil-in-water emulsion and abbreviated o/w) or water droplets dispersed in a continuous oil phase (w/o). These simple emulsions are widely used as lotions and creams for the application of medicinal substances to the skin. They are also used internally, by the oral and parenteral routes, particularly for the delivery of hydrophobic drugs.

A multiple emulsion is defined as consisting of at least three distinct phases. It is possible that such a combination would involve three separate materials that are mutually immiscible. However, pharmaceutically important multiple emulsions developed to date consist of a dispersion in which a phase similar in composition to the continuous liquid medium is itself entrapped. For example, a water-in-oil-in-water (w/o/w) multiple emulsion consists of droplets of a dispersion of water in oil in an aqueous medium. Similarly, it is possible to have a multiple emulsion in which droplets consisting of an o/w dispersion are themselves suspended in oil.

The structure of multiple emulsions may be more easily understood by referring to Figure 1, in which the various phases are pointed out. It is apparent that the identification and naming of the phases in these systems becomes complicated. Sheppard and Tcheurekdjian[1] proposed a naming system capable of handling any multiple-emulsion system likely to be encountered. The use of numbers was proposed to uniquely define the individual components as well as their source in addition to an order terminology to identify the system. Using the example of a w/o/w emulsion, the internal aqueous phase (the water that is dispersed within the oil droplets) would be identified as w1, the oil phase as o, and the external aqueous phase as w2 (Figure 1A). The final descriptive symbolism would be w1/o/w2. Systems with more than one oil phase could be similarly treated. The w1/o/w2 emulsion described above is considered a second-order system as a result of two liquid-liquid interfaces, since each slash represents an interface. If oil makes up the innermost and outermost phases of a second-order system, it would be labeled as an o1/w/o2 emulsion (Figure 1B). This naming convention can be expanded to handle more complex systems. It is useful principally because it permits concise identification of every phase within the emulstion.

The innermost phase (for example, the w1 phase in a w1/o/w2 emulsion) can adopt several configurations.[2] It can exist as a single globule (Figure 1Aa), a group of deflocculated globules (Figure 1Ab), or a cluster of flocculated globules (Figure 1Ac). Configurations may coexist or one of them may predominate, depending on the particular composition employed. Similar possibilities exist for oil-in-water-in-oil (o/w/o) systems. Figure 2 is a photomicrograph of a multiple emulsion.

It is convenient to name the interfaces, beginning from the inside and working outward. Interface I (Figure 1) is the innermost interface. Interface II is the other interface in the simplest multiple emulsion; it separates the medium from the material coating the dispersed globules. In some cases, it is possible to identify the surfactants that stabilize each of these interfaces. Under such conditions, the surfactants would be named accordingly. For example, "surfactant I" would be the stabilizer for interface I.

It is apparent that multiple emulsions have a number of distinct interfaces and large interfacial areas which are responsible for the lack of thermodynamic stability. Mechanisms of destabilization include[2]

1. Coalescence of internal phase droplets
2. Coalescence of dispersed phase globules

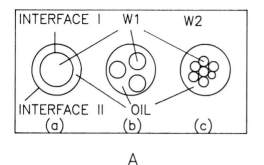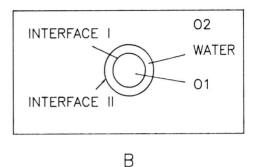

FIGURE 1. Schematic representation of multiple emulsions. (A) w/o/w systems: (a) single internal droplet, (b) deflocculated multiple internal droplets, (c) flocculated multiple internal droplets. (B) o/w/o system. See text for further explanation.

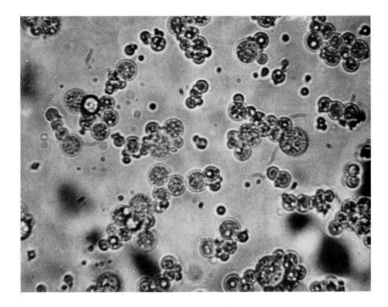

FIGURE 2. Photomicrograph of a dilute w/o/w multiple emulsion system. (Supplied through the courtesy of Dr. Sylvan Frank, College of Pharmacy, The Ohio State University, Columbus.)

3. Disruption of immiscible film separating inner and outer phases
4. Expulsion of internal phase droplets into external phase
5. Transport of material from one phase to another

The presence of an immiscible layer separating inner and outer phases of similar composition is the basis of the utility of the multiple-emulsion system. Multiple emulsions can be used as drug delivery systems or receptor sinks. The ability of the immiscible phase to act as a rate-limiting barrier to the transport of solute determines the ultimate usefulness. Optimization of the system by the introduction of an osmotic pressure differential across the immiscible layer, a change in immiscible phase viscosity, or layer thickness will contribute to modifying the transport properties.

Multiple-emulsion systems have limitations. Foremost is the inherent instability of emulsion systems. Creation of an excessive osmotic pressure differential may destabilize a system by excessive water flux, leading to a complete loss of entrapped water or rupture of the oil layer. Additional difficulties include compositional restrictions caused by the intended use.

Three methods can be used to form multiple emulsions. Combination of all components in specific proportions followed by simple mixing can result in spontaneous multiple-emulsion formation. These systems are of limited usefulness because of compositional and thermal requirements of the system and inability to selectively entrap solutes. Ng and Frank[3] have reported the spontaneous formation of w/o/w and o/w/o multiple emulsions in systems consisting of water, light mineral oil, sorbitan mono-oleate, and polyoxyethylene (20) sorbitan mono-oleate with gentle agitation, but found the regions of stability to change with temperature. Compositional variation was capable of altering the type of emulsion formed. The process of emulsion inversion from w/o to o/w and vice versa has formed multiple emulsions.[4,5] Matsumoto studied the formation of w/o/w multiple emulsions during the inversion of concentrated w/o emulsions.[6] The inversion approach is limited by inflexible composition and the potential of significant mixing of the internal and external phases during manufacture.

The technique most frequently used is the multistep approach where a w/o or o/w emulsion (primary emulsion) is formed and then dispersed in the continuous phase. The result is a multiple emulsion. The advantages of this manufacturing method when compared to others include greater control and flexibility in composition and ease of production. The principal manufacturing difficulty is minimization of miscible phase mixing. This mixing influences the yield of multiple droplets and can result in dispersed droplets which have few, if any,

The term multiple emulsion has been applied to pharmaceutical systems and liquid membrane separatory systems in engineering. There are, however, substantial differences between these systems. Generally, pharmaceutical multiple emulsions have small globules containing dispersed internal droplets. In contrast, liquid membrane separatory systems generally have larger dispersed droplets. The primary emulsion of either system contains emulsifiers, but the liquid membrane system external phase has no emulsifying agents present, whereas the multiple-emulsion system usually has one or more emulsifiers present. This results in differences in properties and uses. The separatory systems do not require long-term stability because their use is generally for short periods of time with destruction being the basis for separation and recovery. These systems also require constant agitation to maintain their disperse nature. The pharmaceutical multiple-emulsion systems, in contrast, require a satisfactory shelf-life without continuous agitation and long-term stability without separation.

Pharmaceutical uses of multiple emulsions have included parenteral chemotherapeutic delivery,[7-11] radiation protection,[12] administration of an antigen adjuvant,[13,14] oral delivery,[15] and removal of endogenous and exogenous toxic materials.[16-19] Table 1 summarizes selected examples of multiple emulsions that have been used in vivo.

II. PREPARATION

Manufacture of multiple emulsions for use as drug delivery systems is limited to the two-stage process described earlier. Yield is defined as the ability to selectively entrap the drug compound in the internal aqueous droplets. Other definitions of yield, in terms of the formation of multiple droplets and their quantity, have been used for studying breakdown characteristics of these systems.

Many factors influence the yield of multiple emulsions. Among the most important are (1) the manufacturing procedure, (2) the component phase volumes, (3) the oil used, and (4) the selection and concentration of the surfactants used. Additional components such as electrolytes or polymers may strongly influence stability; however, their effect is difficult to predict and is dependent on the particular system. Table 2 lists materials used in some emulsion systems that have been mentioned in this chapter.

Table 1
SELECTED EXAMPLES OF MULTIPLE EMULSIONS USED IN VIVO

Drug	Species	Use	Ref.
Drug delivery			
5-Fluorouracil	Rat	Localize drug, prolong release	7
Methotrexate, cytosine arabinoside	Mice, rat	Enhance antitumor activity	8
Bleomycin	Rat	Prolong survival time	9
Bleomycin, mitomycin	Human	Improve activity, reduce administered dose	9
Bleomycin	Rat, rabbit	Prolong absorption, reduce plasma concentrations	10
Cysteamine	Mice	Improve radiation protection	12
Ovalbumin	Mice	Improve antibody titer	13
Influenza vaccine	Human	Improve antibody titer	14
Insulin	Rabbit	Improve oral absorption	15
Biological detoxification			
Quinine sulfate	Rabbit	Diminish absorption	17
Urea	Dog	Decrease systemic urea levels	18
Salicylic acid	Rabbit	Diminish absorption	19

A. Manufacturing Procedure

Two-stage manufacture of a w1/o/w2 emulsion for delivery of a medicinal agent would be completed as follows. In stage 1 the internal aqueous phase, w1, with any additives, is mixed into the oil phase containing surfactant I. Surfactant I stabilizes the w1/o emulsion and produces what is frequently referred to as the primary emulsion. In the second stage, the primary emulsion is added, with mixing, to the w2 phase containing surfactant II to form the final w1/o/w2 emulsion. Surfactant II would be any surfactant or surfactant blend capable of promoting breakup of the w1/o emulsion into globules and stabilization of the final w1/o/w2 system. Manufacture of a multiple emulsion by this method permits the composition of w1 and w2 to be completely different; however, they are most often similar. Because emulsion formation is not quantitative, some mixing of internal and external phases during manufacture must be anticipated.

The energy input of the mixing equipment used to produce a multiple emulsion will have a substantial effect on the yield of final product. Mixing requirements for the first and second stage differ significantly. Production of the primary emulsion (w1/o) requires a high-energy input and turbulent mixing to produce a primary emulsion which has a small dispersed droplet size and a narrow size distribution. Second-stage mixing requires energy sufficient to break the primary emulsion into small, uniformly sized globules, but not great enough to cause rupture of the oil layer and mixing of the w1 and w2 phases. Any type of mixer meeting these requirements is suitable. Continuous mild agitation after manufacture has appeared to offer stability against coalescence.[11,19]

Davis and Walker[20] demonstrated that extent of marker entrapment within the internal phase of a w1/o/w2 emulsion was inversely proportional to the second-stage mixing time. Similar results were found by Magdassi et al.[21] Use of longer sonication times in the second stage of multiple-emulsion formation resulted in smaller-sized multiple droplets, and after 28 days of storage the system sonicated longest had no multiple droplets present.[22] Kondo et al.[23] and Takahashi et al.,[24] working with liquid membrane systems, reported a linear relation between w/o globule breakage and mixing times and mixer speeds.

B. Phase Volumes

Two phase volume ratios are associated wtih a multiple-emulsion system. The first, the primary phase volume ratio, $\phi_{w1/o}$ or $\phi_{o1/w}$, is the ratio of internal phase volume to the total

Table 2
PRINCIPAL COMPONENTS OF REVIEWED
MULTIPLE EMULSION SYSTEMS

Oil	Surfactant	Ref.
Isopropyl myristate	I: A	2
	II: B, C, A/D	
Sesame, peanut	I: A, Q	11
	II: R, S	
Mineral	I: Q	19
	II: T	
Mineral, vegetable	I: A/D	20
	II: D	
Light mineral	I: E	21, 30
	II: D/G	
Mineral	I: A	25
	II: D, K, L, M, N, P	
Aliphatic hydrocarbons	I: A	26
	II: D	
Light mineral	I: E, A/F	28
	II: D/G, A/D, B/U, E/L, A/D	
Mineral	I: E, A/F, F/G	29
	II: D/G, H, D/G/H, D/H/J	

Note: Roman numeral refers to surfactant location: I, surfactant I; II, surfactant II. Slash (/) indicates combination of surfactants. Letters correspond to surfactants listed: A, sorbitan mono-oleate; B, polyoxyethylene (4) lauryl ether; C, polyoxyethylene (16.5) octyl phenol; D, polyoxyethylene (20) sorbitan mono-oleate; E, polyoxyethylene (2) oleyl ether; F, sorbitan trioleate; G, sorbitan monolaurate; H, triethanolamine oleate; J, polyoxyethylene (20) sorbitan trioleate; K, polyoxyethylene (20) sorbitan monolaurate; L, polyoxyethylene (10) oleyl ether; M, polyoxyethylene (9) lauryl ether; N, polyoxyethylene (10) monolaurate; P, polyoxyethylene (10) nonyl phenyl ether; Q, sorbitan sesquioleate; R, polyoxyethylene (10) hydrogenated castor oil; S, polyoxyethylene (60) hydrogenated castor oil; T, cetyltrimethylammonium bromide; and U, polyoxyethylene (23) lauryl ether.

volume of the primary emulsion (the internal phase is indicated on the left side of the slash). The secondary phase volume ratio, $\phi_{w1/o/w2}$ or $\phi_{o1/w/o2}$, is the ratio of volume of the primary emulsion to the total volume of multiple emulsion. These ratios have also been described by terms such as "primary" or "secondary volume fraction". Multiplication of the primary phase volume ratio by the secondary phase volume ratio yields the fraction of internal phase. Multiplication of the quantity $(1 - \phi_{w1/o})$ by the secondary phase volume ratio provides the fraction of immiscible phase. The external volume fraction is determined by difference.

Matsumoto et al.[25] found that changes in the primary phase volume ratio had little influence on the yield of several water/hydrocarbon oil/water multiple emulsions studied. In this study the primary phase volume ratio ranged from 0.3 to 0.9 and 95 to 98% of the glucose marker was entrapped. Davis and Walker[20] reported substantially different results. In a number of systems utilizing different oils, they reported that an increase in the primary phase volume ratio produced a resultant decrease in percent entrapment of the marker 6-carboxyfluorescein. The resultant curves of primary phase volume ratio vs. percent entrapment were approximately sigmoidal. These studies report conflicting results which may be related to the substantially different concentration of surfactant I used.

Davis and Walker[20] showed that the effect of an increase in the secondary phase volume ratio depended on the oil. With squalane, the percent entrapment increased dramatically with increases in the secondary phase volume ratio and then approached a maximal value, while light mineral oil, maize, sesame, and peanut oil showed little change in entrapment percent at any phase volume ratio.

Coalescence between inner aqueous droplets and outer aqueous phase took place more frequently as the volume of inner aqueous phase was increased.[11] The thickness of the oil layer was not influenced by the primary phase volume.[26] From these findings it was concluded that the oil layer immediately thinned after the second step of emulsification, with the surplus oil appearing heterogeneously on the oil layers or dispersing into the continuous medium as oil droplets. This conclusion is supported by the work of Takahashi et al.,[24] who found that immiscible film disruption was not affected by primary phase volume ratio.

C. Oil Influence

The selection of the oil phase component must be based on the intended use of the multiple-emulsion system. In the case of a topical or oral system, there would be many oils to choose from, but if parenteral administration is intended, the choice is much more limited. Metabolizable oils, such as vegetable oils, are appropriate for this application. It must be remembered that the oil selection can have a marked effect on the amount of solute entrapment and the stability of a multiple emulsion.

Multiple emulsions formed more easily when they contained hydrocarbon oils rather than vegetable oils.[27] Davis and Walker[20] studied multiple-emulsion systems containing different oils and reported several interesting findings. The following order for the yield, based on entrapment efficiency, was reported: light liquid paraffin > squalane > sesame oil > arachis (peanut) oil > maize oil. The oils used were of high purity; however, the authors noted that vegetable oils could contain small amounts of surface-active materials (such as monoglycerides) which might compete with surfactants for space at the interface during primary emulsion formation. No correlation between the physical properties of the various oils used and the yield of multiple-emulsion droplets was found. Fukushima et al.[11] compared sesame oil and peanut oil for use in a parenteral application. Peanut oil was selected for use because during refrigeration a stability difference was observed which the authors related to the content of saturated fatty acids in the oils.

Water permeation of the immiscible layer is frequently utilized as an indicator of system stability and is indicative of transport potential. The oil selected can affect this permeation coefficient. The water permeation coefficient at a constant osmotic pressure gradient was reported to decrease slightly as the carbon number of the hydrocarbon making up the oil phase increased.[26] The rate of swelling of the vesicles under the osmotic pressure gradient was inversely related to oil viscosity.

D. Surfactant Selection

No other single factor has as great an influence on the production of stable multiple-emulsion systems as does the selection of the surfactant system. Surfactants are needed for multiple-emulsion formation and it is their interaction with the oil and the aqueous phase which determines ultimate stability. Several investigators have examined and reported on the influence of the surfactants used on the yield and stability of multiple emulsions.[2,25,28]

Nonionic surfactants have been most frequently used; however, infrequent reports exist for the use of charged surfactants.[29] Matsumoto et al.[25] reported that nonionic emulsifying agents are preferable to all ionic ones for the second emulsification in order to achieve maximal entrapment efficiency. Also, with nonionic surfactants there is less potential for irritation to tissues or charge interactions with drugs.

The hydrophile-lipophile balance (HLB) concept is useful in emulsifier selection but, as in simple emulsion systems, it provides only a starting point in the selection of emulsifiers for a multiple-emulsion system. A surfactant having a low HLB number favors the formation of a w/o emulsion. This emulsion would become the primary emulsion of a w/o/w multiple system. Optimization of the required HLB for the w1/o emulsion can be accomplished in the usual manner. A high HLB number surfactant (itself favoring formation of an o/w emulsion) will assist in the breakup and dispersion of the w1/o emulsion and result in a multiple-emulsion system. Conversely, use of a high HLB surfactant as surfactant I and a low HLB surfactant as surfactant II will result in an o/w/o multiple emulsion.

Florence and Whitehill[2] have shown that the surfactant selected can determine the configuration of the multiple system produced. Using various nonionic surfactants as surfactant II in fixed concentrations and a fixed concentration of surfactant I, three distinct droplet types were observed. These droplet types did not exist exclusively in a particular system, but usually predominated so identification of the system could be made. For example, the droplets formed with polyoxyethylene (4) lauryl ether were relatively small and contained usually one and at most a few relatively large internal drops. Those emulsions formed with polyoxyethylene (16.5) octyl phenol consisted of larger multiple globules and contained smaller but more numerous internal droplets. Emulsions formed with a blend of sorbitan mono-oleate and polyoxyethylene (20) sorbitan mono-oleate in the ratio of 3:1 were the most complex, consisting of large numbers of internal aqueous drops which were too numerous to count.

Multiple-emulsion formation was influenced by the concentration of surfactant I used in the first stage of emulsification.[25] For a fixed concentration of surfactant II and fixed phase volume ratios, it was found that more than 30% by weight of sorbitan mono-oleate as surfactant I was required in the oil phase to obtain better than 90% entrapment of marker. Similar results were found for sorbitan monolaurate. Liquid membrane systems which lack the presence of surfactant II have been shown to form stable liquid membrane systems with surfactant I concentrations in the range of 1 to 2%.[23,24] This would indicate the concentration of emulsifiers I and II and their ratio to be significant factors in the production of stable systems.

The concentration of surfactant II has been shown to influence multiple-emulsion stability.[21,25,29,30] Multiple-emulsion yield decreased as the concentration of hydrophilic surfactant was increased.[21,25,29] Magdassi et al.[21] suggested that very low concentrations of emulsifier II would result in a w1/o/w2 system similar to a liquid membrane system in which the yield of preparation does not change significantly at any HLB. Microscopic observations demonstrated the existence of large, irregularly shaped globules of primary emulsion and concluded that only mechanical rupture was determining the emulsion yield. At higher emulsifier II concentrations, there was a continuous decrease in mulitple-emulsion droplet size, accompanied by a decrease in marker entrapment.

Yield reduction with increasing surfactant II concentrations was attributed,[25] to surfactant I migration to the external interface and loss into the continuous phase by surfactant II solubilization. Garti and colleagues[21,29] associated the decreased yield to formation of smaller droplets with an increased amount of inner droplet breakage. An investigation[30] into this phenomenon revealed a shift in the required system HLB caused by migration of emulsifiers. Frenkel et al.[28] reported that the "weighted" HLB value, which was the HLB value calculated from all the emulsifiers present, was important in determining the point at which a w/o/w multiple emulsion inverted to an o/w emulsion.

The choice of surfactants in a multiple-emulsion system can have a significant effect on the observed properties, some of which are difficult to predict. Chilamkurti and Rhodes[31] reported that the surfactant chosen to stabilize a liquid membrane system had a substantial effect on the solvent properties of the oil phase when comparing apparent partition coefficients in the oil and the liquid membrane system. The high concentration of oil-soluble surfactants present in many of these systems would be expected to facilitate passage of materials by solubilization.

Table 3
PRINCIPAL TECHNIQUES USED IN ASSESSING MULTIPLE EMULSIONS

Technique	Property measured	Ref.
Microscopy	Droplet size distributions of internal drop-lets and primary emulsion globules Dispersion pattern of internal droplets Immiscible phase layer thickness	2, 22, 25, 26, 29, 32
Coulter Counter®	Droplet size distribution of primary emul-sion globules	22
Rheology	Empirical measure of volume fraction of the primary emulsion	33—37
Tracer studies	Permeability and/or disruption of the immis-cible phase layer	10, 11, 19, 20, 25, 29, 31, 32, 38-43

The various factors which influence the formation of a multiple-emulsion system cannot be considered separate from one another. The parameters which have been reviewed have been selected from various studies with the intention of considering a particular effect. The observed properties in any actual system will be a function of the interrelationship between factors.

III. ASSESSMENT

Multiple emulsions have unique characteristics which must be considered before these systems can be used for drug delivery. Considerable effort has focused on studying emulsion structure and physical stability, in addition to determining uptake or release kinetics of these systems.

The complex structure of a multiple emulsion necessitates studying two types of droplets, those consisting of w1 or o1, the encapsulated internal phase, and the primary emulsion, either w1/o or o1/w in the w2 or o2 phase, respectively (see Figure 1). It is important to be able to determine the proportion of multiple droplets (w1/o/w2 or o1/w/o2) to simple emulsion droplets (w/o or o/w) present. Where possible, it is advantageous to obtain the respective droplet size distributions and to know how these change with respect to time. The principal techniques that have been used for assessment are summarized in Table 3. They are described in the following paragraphs.

A. Microscopy

Microscopy has been used by a number of investigators to document the existence of a multiple emulsion prior to experimentation as well as to study properties of these systems.[2,22,29,32] Investigators[2,26] have used microscopy to estimate the water permeability across the immiscible layer due to an osmotic gradient in a w1/o/w2 emulsion system.

Florence and Whitehill[2] studied the breakdown features of a water/isopropyl myristate/water multiple emulsion by photomicrography and cinematography. An osmotic gradient was utilized to induce water flow across the oil layer. The trans-oil layer water permeability was calculated from a photographic estimation of the average w1/o globule diameter and a photographic sequence of the globule volume shrinkage caused by the osmotic gradient. It was assumed that true osmotic flow occurred or, in other words, that the oil was impermeable to the solute used to create the gradient, but permeable to solvent. (The potential for error using this assumption will be discussed in Section III.C.) The value obtained from the system studied compared favorably to that found by Matsumoto and Kohda[33] using a viscometric technique.

Matsumoto et al.[26] measured the swelling rate of w1 droplets in w1/o/w2 multiple emulsions under an osmotic pressure gradient and photomicrographically recorded the process. Swelling resulted from the migration of water from the w2 phase to the w1 phase across the intervening oil layer because solute was initially placed in the internal phase. Microscopy permitted determination of the rate of volume flux of water and the surface area of the globule at any time. These investigators could then estimate the water permeation coefficient of the oil layer in a w1/o/w2 emulsion system.

Microscopy has been used as an aid in determining the thickness of the immiscible layer. Matsumoto et al.[25] assumed a single internal aqueous drop surrounded by an oil phase of uniform thickness, calculated an average oil layer thickness, and attempted to verify this estimation visually. Visualization was hindered by the similar refractive indexes of the oil and aqueous phases. However, Florence and Whitehill[2] and Davis and Burbage[22] demonstrated the possibility of producing multiple emulsions containing numerous internal aqueous droplets without densely packing the w1/o globules. In such systems the analysis of Matsumoto et al.[25] might be inappropriate.

Sizing of the dispersed internal droplets has provided the greatest challenge. These droplets are small and can approach the resolution limits of a light microscope. Brownian motion and circulation within the immiscible phase droplet complicate clear visualization. Difficulty in discriminating the interface between the single internal aqueous droplet and surrounding oil film in a w1/o/w2 emulsion has been reported.[25,26]

Davis and Burbage[22] observed a bimodal size distribution in a w1/o/w2 multiple emulsion because some globules contained internal droplets while others did not. They resolved the composite distribution into components of w1/o and simple oil globules by use of a graphical inflection technique. These authors also reported using a freeze-etching technique to produce electron micrographs in an effort to study the internal aqueous droplet size and distribution. However, the technique is limited in application by the cost, time required for preparation, and large number of micrographs needed for statistical analysis.

B. Coulter Counter®

Davis and Burbage[22] utilized the Coulter Counter®, as well as microscopy, to study the characteristics of the w1/o droplets of a w1/o/w2 multiple emulsion. The water flux across the oil layer was induced by an osmotic gradient and detected by the instrument as a shrinkage of the dispersed globule. The bimodal globule size distribution initially present was transformed to a simple log-normal distribution by osmosis. The proportion of large droplets containing dispersed water droplets decreased while the proportion of small droplets consisting of oil increased with time. Microscopy confirmed that the final system consisted of a simple o/w emulsion.

The multiple emulsion studied[22] contained many individual internal droplets (similar to Figure 1Ab and c). However, the mathematical treatment developed assumed the presence of a single large internal aqueous droplet, because water diffusion through the oil layer was thought to be controlled only by the outermost internal aqueous droplets.

The objective of this technique and analysis was to permit use of an instrumental method to study storage stability. As an example, it was rapidly determined that the mean volume diameter of a water/light mineral oil/water emulsion decreased approximately 6% during 60 days. During the same time period, the volume of the internal water phase decreased 27%.

C. Rheology

The volume of multiple-emulsion droplets (e.g., w1 phase in a w1/o/w2 emulsion) is capable of changing because of flow of solvent (e.g., water) across the oil layer or its rupture. To study these phenomena, it is necessary to establish a quantitative relationship between the droplet volume fraction and some measurable property. Because direct measurement of changes in droplet volume may not be possible, indirect means such as viscometry are often used. Multiple-emulsion stability[34,35] and oil layer permeability to water[33] and solutes[36] have been studied using this technique.

Kita et al.[34] studied w1/o/w2 multiple emulsions consisting of a single inner aqueous droplet dispersed in the oil globule. The volume fraction of primary emulsion (w1/o globules) was kept low to maintain Newtonian flow characteristics, a requirement for the analysis, which was based on Mooney's equation (Equation 1):

$$\ln \eta_{REL} = \alpha\phi_T/(1 - \lambda\phi_T) \tag{1}$$

The relative viscosity (η_{REL}) was related to the volume fraction of the dispersed phase (ϕ_T), the shape factor of the dispersed globules (α), and the crowding factor of the globules (λ). This equation provided a relationship between the measurable relative viscosity and the encapsulated aqueous phase volume. The constants α and λ depended on the system of investigation.

Matsumoto and Kohda used a viscometric method in an attempt to estimate the water permeation coefficient of the oil layer in a w1/o/w2 system placed in an osmotic gradient.[33] It was possible from their analysis to resolve the changes in internal water phase volume ratio into an osmotic pressure contribution and the effect of solubilization of water into the oil phase.

Matsumoto et al.[35] attempted to estimate the stability of the oil layer in a w1/o/w2 emulsion by viscometry. A series of emulsions containing various hydrocarbon oils were found stable for at least 1 month. The rate of volume flux of water across the oil layer and the total area of the oil layer in a unit volume of emulsion for a given oil remained relatively constant over the 30-day observation period.

The technique of determining internal phase volume by viscometric assessment requires the system to be Newtonian, making it mandatory that the phase volume of the primary emulsion be kept small and that relatively low shear rates be used in testing. Thus, viscometric analysis, while applicable to systems of theoretical interest, cannot be directly applied to multiple emulsions with large internal phase volumes required for drug entrapment and is therefore limited in its application to drug delivery systems.

Tomita et al.[36] utilized low shear viscometry on a non-Newtonian system to qualitatively measure solute permeability through the oil layer. It was shown that water as well as various solutes were capable of passing through the oil layer. The reported permeability sequence for the solutes tested was urea > potassium thiocyanate > potassium chloride > glucose > calcium chloride. The authors used these results to conclude that a study by Matsumoto et al.[25] which monitored the appearance of glucose as a measure of entrapment efficiency and emulsion stability would give incorrect results because of the ability of glucose to pass the oil layer.

Matsumoto and Sherman[37] studied a water/olive oil/water multiple emulsion viscometrically and a component surfactant film via creep compliance. It was found that addition of the saccharides glucose and sucrose caused the system to change viscosity in accordance with changes resulting from true osmotic flow; however, the electrolytes and organic acids used caused nonideal viscometric changes. The authors concluded that the presence of ions could cause changes in the oil layer or the interfaces, which necessitates considering the oil layer and two interfacial regions as a single structure.

D. Tracer Studies

Several requirements must be satisfied before a material can be considered for use as a marker. First, knowledge of the mode of transport across the immiscible phase is required. Substances which diffuse across the immiscible layer can be used to determine time-dependent release or uptake characteristics of a particular system, as well as the effects of different additives or manufacturing procedures. Substances which are unable to diffuse can be used to assess the vulnerability of the system to breakage or separation. An additional requirement of any inert material added to study system stability is that it should not itself affect the overall stability of the system.

Several classes of materials have been used as tracers. Electrolytes have been frequently used to monitor the stability of the immiscible layer in w1/o/w2 systems, in addition to providing an osmotic pressure gradient. Most frequently, electrolytes are formulated into the internal aqueous phase. Medicinal agents[10,11,29,32,38] and inert nonelectrolytes[20,25,39] have been used to assess release characteristics. Uptake kinetics into the internal aqueous phase droplets has been studied using drug compounds.[16,19,31]

Radiotracing using tritiated water has been effectively utilized as a nonperturbative technique to monitor water exchange across the oil membrane in w1/o/w2 systems.[40] The primary advantage of this technique is the absence of an induced osmotic pressure gradient and complications caused by solute addition to the emulsion. Pilman et al.[39] used release of tritiated water from the internal aqueous phase into a dialysis sink to assist in studying transport of water-soluble materials in inverse micellar multiple emulsions stabilized by different proteins.

Direct determination of the tracer concentration in the external continuous phase is seldom practical. Separation of the primary emulsion from the external phase is necessary when direct determination of a solute in the intact system is not possible. The optimum separatory procedure is dependent on the tracer chosen and the experimental information desired. Dialysis[11,19,25,29,32,38,41] and filtration[20,31,42,43] methods are the most common separatory procedures.

Dialysis may cause the system to experience an unanticipated osmotic pressure gradient or adverse dilutional effects, while filtration may cause breakage of the primary emulsion. Only prior experimentation will determine what artifacts, if any, are introduced.

Pilman et al.[39] kept dialysis periods short to prevent dilutional effects from causing unwanted phase transitions in the systems being studied. Davis and Walker[20] used a filtration procedure to effect separation of primary emulsion from continuous phase. Their interest was in monitoring the release of a tracer dye from the w1 to w2 phase. The separation was accomplished by diluting a sample to produce creaming followed by subsequent filtration through a micropore membrane. These investigators reported that this procedure did not disrupt primary emulsion droplets.

Magdassi and Garti[44] have proved that simple electrolytes used as markers can diffuse from the w1 to the w2 phase in a w/o/w multiple emulsion. Their work confirms the conclusions of Tomita et al.[36] whose studies provided indirect evidence of this behavior. Since low molecular weight salts do not necessarily remain associated with an entrapped aqueous phase, their use as markers to indicate stability is questionable.

Investigations in our laboratory have resulted in development of a stability-indicating technique which uses a high molecular weight polymeric dye as a marker. The dye is essentially insoluble in neat mineral oil and in oil containing emulsifiers, nor does it partition into mineral oil from an aqueous phase. If the dye is incorporated into the internal aqueous phase (w1) of a multiple emulsion, its appearance in the external aqueous phase (w2) is an indication of breakage of the oil film separating these phases, since diffusional interphase transport does not occur.

The dye marker (polyporphyre) was dissolved in the aqueous liquid intended to be the w1 phase and the two-stage manufacturing procedure was used to prepare a series of mulitple emulsions. Analysis involved addition of a second dye to an aliquot of the emulsion as external standard. Gentle centrifugation was employed to separate the external aqueous (w2) phase from the rest of the emulsion system. Complete clarification was accomplished by addition of tetrahydrofuran immediately followed by spectrophotometric analysis of the

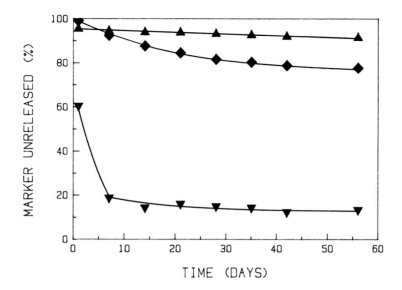

FIGURE 3. Retention of polymeric marker in w1 phase as a function of time in multiple emulsions containing various concentrations of surfactant II; (▲) = 0.33%, (◆) = 0.95%, and (▼) = 2.38%.

solution at two wavelengths. By solving a set of simultaneous equations, the amount of marker dye present in the w2 phase was calculated. It is important to know the amount of marker rather than its concentration since the latter may be influenced by diffusional transfer of water between the internal and external aqueous phase. The amount of marker appearing in the w2 phase provides a direct measure of oil film rupture in the emulsion.

Preliminary results for a group of emulsions made with the same concentration of surfactant I (sorbitan mono-oleate and polysorbate 80; HLB = 6.0 in the primary emulsion) and different amounts of surfactant II (polysorbate 80 in the w2 phase) are presented in Figure 3. For these systems, increasing the concentration of surfactant II resulted in an increase in the rate of dye release. In the most stable of the three emulsions, more than 90% of the dye remained entrapped in the w2 phase for nearly 2 months.

E. Other Methods

Davis[45] reported that loss of internal aqueous droplets in a multiple emulsion caused the droplet interior to change appearance from dark grey to clear and proposed the possibility of using a photodensiometric technique to follow changes in globule size. No further investigations of this phenomenon have been reported.

IV. TRANSPORT PHENOMENA

Several mechanisms for transfer between the internal and external phases of multiple emulsions have been identified. In a w1/o/w2 emulsion, disruption of the oil phase or digestion of the oil will release all of the material contained within the w1 phase. This may occur, for example, when a multiple emulsion containing a digestible oil is administered orally. If the oil layer remains intact, then drug transfer can take place by simple diffusion along a concentration gradient. Other mechanisms, such as micellar transport and carrier-mediated transport, are also possible.

A. Diffusional Transport

Simple diffusion or passive transport is the principal means of transfer across the immiscible layer in most pharmaceutical multiple-emulsion systems. The flux of material crossing the immiscible layer is dependent on permeant properties (such as pK_a and partition coefficient) and emulsion properties, including the type and concentration of surfactants, nature and concentration of any electrolytes present, and thickness and viscosity of the intervening layer.

Tomita et al.[36] have reported the ability of solutes (both electrolytes and nonelectrolytes) to pass through an oil layer in a w1/o/w2 emulsion and a planar membrane system. Changes in relative viscosity were used to document solute flux in a multiple emulsion while direct monitoring of flux was accomplished in a diffusion cell divided by a planar membrane which had been soaked in the hydrophobic surfactant.

Several workers have studied multiple emulsions of the w1/o/w2 type in which the w1 phase acted as a sink for the uptake of drug materials. The application was intended for detoxification. Morimoto et al.[19] monitored drug appearance across a cellulose membrane and into a receptor compartment. The donor compartment contained free drug either with or without multiple emulsion. Diminished drug concentrations were observed in the receptor compartment when multiple emulsions were present with the free drug, indicating emulsion uptake. In vivo detoxification of salicylic acid in rabbits was measured by monitoring blood salicylate levels. Effective detoxification was demonstrated by the tested systems.

Rhodes and co-workers[31,42,43] investigated several w1/o/w2 systems in vitro. From these studies, attempts were made to explain drug transport according to Fickian diffusion theory. A constant concentration gradient was maintained by buffering the pH on either side of the immiscible membrane to different values. In these studies the external aqueous phase was strongly acidic and the internal aqueous phase was strongly alkaline. The acidic drug was transported across the immiscible oil layer and was ionized at the alkaline internal aqueous pH. Because the immiscible layer was oil, passage of the ionized species back out of the internal phase did not occur. Drug uptake was monitored by observing the concentration loss from the external aqueous phase.

Chiang et al.[42] found drug transport to obey first-order kinetics and follow Fickian diffusion theory. Using the internal phase of a liquid membrane system as a receptor sink, theoretical transport of nonionized drug across the membrane can be expressed (see Figure 4) as:

$$\text{rate} = dC_0/dt = -DA(\Delta C/\Delta X) = -(DA/\Delta X)(C_0 - C_1) \tag{2}$$

where D is the diffusion coefficient of drug in the oil layer, ΔC is the concentration differential across the oil layer, A is the area of the w1/o globule, and ΔX is the thickness of the oil layer. C_0 and C_1 are concentrations within the oil at the outer and inner boundaries, respectively. The drug concentration of the internal aqueous phase (receptor) is labeled 1. Using the partition coefficient relationship $P = C_0/C_2$, where C_2 is the concentration in the donor phase, assuming sink conditions, and integrating leads to

$$C_2 = C_{20} \exp(-DAt/\Delta X) \tag{3}$$

where C_{20} is the initial concentration of drug in the external phase.

Yang and Rhodes[43] presented a simple model to describe the transport of solutes across a liquid membrane:

$$C \underset{k_{21}}{\overset{k_{12}}{\rightleftharpoons}} C_{memb} \underset{k_{32}}{\overset{k_{23}}{\rightleftharpoons}} C_2 \tag{4}$$

where the C and C_2 have the same meaning as above; C_{memb} is the drug concentration in

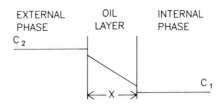

FIGURE 4.   Concentration diagram for diffusional transfer from external phase (phase 2) to internal phase (phase 1) of a multiple emulsion.

the liquid membrane layer, and k_{12}, k_{21}, k_{23}, and k_{32} are the first-order microrate constants between the three compartments. Therefore the concentrations of solute in the external phase at any time t, C_t, is given by

$$C_t = A \exp(-\alpha t) + B \exp(-\beta t) + Css \qquad (5)$$

where A and B are preexponential terms, α and β correspond to complex first-order macrorate constants, and Css is the concentration of drug in the external aqueous phase when equilibrium is reached. If

$$\Delta C_t = C_t - Css \qquad (6)$$

then a plot of log ΔC_t as a function of time should be biphasic with slopes of alpha and beta. This model was found to be operative with the alpha constant being attributed to transport of solute into the oil and the beta constant to transport into the internal aqueous phase.

Frank and co-workers[32,41] studied in vitro release of drug compounds from w/o/w and o/w/o multiple-emulsion systems. These investigators reported diffusion coefficients calculated from the slope of the plot of the percentage of released drug as a function of the square root of the time. It was assumed that (1) the tested emulsions constituted a single homogeneous phase and (2) the concentration of solute at time zero was the total concentration of solute, dissolved and suspended, in the system.

The influence of increased external phase viscosity and additive presence in the internal phase was studied in a w/o/w system.[32] It was found that diffusion in the external aqueous phase was rate determining, with the oil layer acting as an encapsulating liquid membrane. Sodium chloride and sorbitol addition to the internal aqueous phase influenced the drug diffusion coefficient, but to differing degrees. The observed diffusion coefficient initially decreased with increasing additive concentration. This was attributed to a change of drug activity with a resultant decreased driving force across the oil layer. Further increases in the additive concentration were proposed to cause a salting out of surfactant present in the internal aqueous phase and resulted in a more effective interfacial barrier and a further lowering of the diffusion coefficient. Still higher additive concentrations caused a further condensation of the interfacial surfactant layer with resultant loss of ability to stabilize the system. The result was diminished emulsion stability and enhanced diffusion.

Explanation of the diffusion coefficient behavior and drug release in an o/w/o system was more complicated.[32,41] Changes in the aqueous sodium chloride concentration were shown to influence the effective diffusion coefficient in a pattern similar to that seen in the w/o/w system above; however, the minimal diffusion coefficient was observed at a significantly lower electrolyte concentration. This was ascribed to the use of a different surfactant in the system and not a differing process. Interface II appeared to be most sensitive to increasing sodium chloride concentration in the o/w/o system.

Location of the drug in the internal or external oil phase of an o/w/o system was shown[32,41] to produce different release patterns. Although the intention was to localize drug in the internal oil phase, drug leakage to the external oil phase occurred during manufacture or on standing. Drug release from the external oil phase into an aqueous sink, identified as the alpha release phase, was shown to be equivalent to release from an oil suspension and to drug localization in the external oil phase. Drug release from the primary emulsion, identified as the beta phase, was observed when drug was depleted from the external oil phase. Release through the ol/w interface was then observed and found to be rate limiting. Appearance of the beta phase could be hastened or inhibited by changing the initial drug concentration in the external oil phase.

Brodin and Frank[41] used the same o/w/o system[32] to study the effects of drug concentration and hydrophobicity on the rate of drug release. They reported a substantially reduced diffusion coefficient for the compound having the greater partition affinity for the oil; however, the drugs used differed markedly in chemical structure. They concluded that, in the case of the system studied, factors other than hydrophobicity were involved in the release process. It was also observed that the selected drug and its concentration had a significant influence on the number of globules containing multiple droplets, indicating a marked influence on the emulsion formation characteristics.

B. Micellar Transport

Multiple emulsions generally contain one or more surfactants in relatively high concentration. Surfactants with low HLB value may form inverse micelles within the oil phase in a w1/o/w2 emulsion (in these micelles, the polar heads are in the center and hydrocarbon tails extend outward). Water and other polar substances can be solubilized within micelles and, in this way, transported across the oil film separating the aqueous phases. In similar fashion, hydrophobic materials can be transported by "normal" micelles of a surfactant with high HLB across the water layer in an ol/w/o2 emulsion.

The water diffusion coefficient across an isopropyl myristate oil layer in a w1/o/w2 multiple-emulsion system was similar to that of the inverse micelles, supporting the idea that water transport across the oil layer was by way of inverse micelles.[2] Chilamkurti and Rhodes[31] suspected inverse micelles to be responsible for transport of some ionized species across oil layers.

C. Carrier-Mediated Transport

Carrier-mediated transport or facilitated transport is frequently utilized in chemical extraction systems, particularly for waste water treatment. The structure of carriers capable of transporting materials is complex. Carrier selection is based on the structure of the molecule to be transported, but selection of a particular carrier appears to be somewhat empirical.[46]

The carrier molecule and the carrier-material complex must be soluble in the immiscible phase, but not the external or encapsulated interior phase. The complex must also be capable of diffusing in the immiscible layer. This results in localization and transporting ability. Complexation must be reversible; separation of carrier and material must be capable of occurring at the interface opposite the one at which complex formation occurs. Elimination of the transported species in the internal phase by ionization, irreversible binding, or precipitation permits transport against a concentration gradient. While theoretically applicable to systems of pharmaceutical interest, this transport method has yet to be successfully employed.

It is apparent that multiple emulsions provide a challenge in formulation and manufacture and that a great deal of fundamental work remains to be done. The number of potential uses and promising preliminary work indicates that the investigative effort is indeed worthwhile.

ACKNOWLEDGMENTS

The authors wish to thank the Kelco Division, Merck and Co., and Roche Laboratories, Hoffman La Roche, Inc., for fellowship support during preparation of this manuscript.

REFERENCES

1. **Sheppard, E. and Tcheurekdjian, N.,** Comments on multiple phase emulsions, *J. Colloid Interface Sci.,* 62, 564, 1977.
2. **Florence, A. T., and Whitehill, D.,** Some features of breakdown in water-in-oil-in-water multiple emulsions, *J. Colloid Interface Sci.,* 79, 243, 1981.
3. **Ng, S. M. and Frank, S. G.,** Formation of multiple emulsions in a four-component system containing nonionic surfactants, *J. Dispersion Sci. Technol.,* 3, 217, 1982.
4. **Seifriz, W.,** Studies in emulsions, *J. Phys. Chem.,* 29, 738, 1925.
5. **Mulley, B. A. and Marland, J. S.,** Multiple-drop formation in emulsions, *J. Pharm. Pharmacol.,* 22, 243, 1970.
6. **Matsumoto, S.,** Development of w/o/w-type dispersion during phase inversion of concentrated w/o emulsions, *J. Colloid Interface Sci.,* 94, 362, 1983.
7. **Takahashi, T., Mizuno, M., Fujita, Y., Ueda, S., Nishioka, B., and Majima, S.,** Increased concentration of anticancer agents in regional lymph nodes by fat emulsions with special reference to chemotherapy of metastasis, *Gann,* 64, 345, 1973.
8. **Benoy, C., Schneider, R., Elson, L., and Jones, M.,** Enhancement of the cancer chemotherapeutic effect of the cell cycle phase specific agents methotrexate and cytosine arabinoside when given as a water-oil-water emulsion, *Eur. J. Cancer,* 10, 27, 1974.
9. **Takahashi, T., Ueda, S., Kono, K., and Majima, S.,** Attempt at local administration of anticancer agents in the form of fat emulstion, *Cancer,* 38, 1507, 1976.
10. **Yoshioka, T., Ikeuchi, K., Hashida, M., Muranishi, S., and Sezaki, H.,** Prolonged release of bleomycin from parenteral gelatin sphere-in-oil-in-water multiple emulsion, *Chem. Pharm. Bull.,* 30, 1408, 1982.
11. **Fukushima, S., Juni, K., and Nakano, M.,** Preparation of and drug release from w/o/w type double emulsions containing anticancer agents, *Chem. Pharm. Bull.,* 31, 4048, 1983.
12. **Gresham, P., Barnett, M., Vaughn-Smith, S., and Schneider, R.,** Use of a sustained release multiple emulsion to extend the period of radioprotection conferred by cysteamine, *Nature (London),* 234, 149, 1971.
13. **Herbert, W. J.,** Multiple emulsions. A new form of mineral oil antigen adjuvant, *Lancet,* 2, 771, 1965.
14. **Taylor, P. J., Miler, C. L., Pollock, T. M., Perkins, F. T., and Westwood, M. A.,** Antibody response and reactions to aqueous influenza vaccine, simple emulsion vaccine and multiple emulsion vaccine. A report to the Medical Research Council Committee on influenza and other respiratory virus vaccines, *J. Hyg. Camb.,* 67, 485, 1969.
15. **Shichiri, M., Shimizu, Y., Yoshida, Y., Kawamori, R., Fukuchi, M., Shigeta, Y., and Abe, H.,** Enteral absorption of water-in-oil-in-water insulin emulsions in rabbits, *Diabetologia,* 10, 317, 1974.
16. **Frankenfeld, J. W., Fuller, G. C., and Rhodes, C. T.,** Potential use of liquid membranes for emergency treatment of drug overdose, *Drug Dev. Commun.,* 2, 405, 1976.
17. **Morimoto, Y., Yamaguchi, Y., and Sugibayashi, K.,** Detoxification capacity of a multiple (w/o/w) emulsion for the treatment of drug overdose. II. Detoxification of quinine sulfate with the emulsion in the gastro-intestinal tract of rabbits, *Chem. Pharm. Bull.,* 30, 2980, 1982.
18. **Asher, W. J., Bovee, K. C., Frankenfeld, J. W., Hamilton, R. W., Henderson, L. W., Holtzapple, P. G., and Li, N. N.,** Liquid membrane system directed toward chronic uremia, *Kidney Int.,* 7, S409, 1975.
19. **Morimoto, Y., Sugibayashi, K., Yamaguchi, Y., and Kato, Y.,** Detoxification capacity of a multiple (w/o/w) emulsion for the treatment of drug overdose: drug extraction into the emulsion in the gastro-intestinal tract of rabbits, *Chem. Pharm. Bull.,* 27, 3188, 1979.
20. **Davis, S. S. and Walker, I.,** Measurement of the yield of multiple emulsion droplets by a fluorescent tracer technique, *Int. J. Pharm.,* 17, 203, 1983.
21. **Magdassi, S., Frenkel, M., and Garti, N.,** On the factors affecting the yield of preparation and stability of multiple emulsions, *J. Dispersion Sci. Technol.,* 5, 49, 1984.
22. **Davis, S. S. and Burbage, A. S.,** The particle size analysis of multiple emulsions (water-in-oil-in-water), in *Particle Size Analysis,* Groves, M. J., Ed., Heyden, London, 1978, 395.

23. **Kondo, K., Kita, K., Koida, I., Irie, J., and Nakashio, F.,** Extraction of copper with liquid surfactant membranes containing benzoylacetone, *J. Chem. Eng. Jpn.,* 12, 203, 1979.
24. **Takahashi, K., Ohtsubo, F., and Takeuchi, H.,** A study of the stability of (w/o)/w-type emulsions using a tracer technique *J. Chem. Eng. Jpn.,* 14, 416, 1981.
25. **Matsumoto, S., Kita, Y., and Yonezawa, D.,** An attempt at preparing water-in-oil-in-water multiple phase emulsions, *J. Colloid Interface Sci.,* 57, 353, 1976.
26. **Matsumoto, S., Inoue, T., Kohda, M., and Ikura, K.,** Water permeability of oil layers in w/o/w emulsions under osmotic pressure gradients, *J. Colloid Interface Sci.,* 77, 555, 1980.
27. **Davis, S. S.,** Liquid membranes and multiple emulsions, *Chem. Ind. (London),* 19, 683, 1981.
28. **Frenkel, M., Shwartz, R., and Garti, N.,** Multiple emulsions. I. Stability: inversion, apparent and weighted HLB, *J. Colloid Interface Sci.,* 94, 174, 1983.
29. **Garti, N., Frenkel, M., and Shwartz, R.,** Multiple emulsions. II. Proposed technique to overcome unpleasant taste of drugs, *J. Dispersion Sci. Technol.,* 4, 237, 1983.
30. **Magdassi, S., Frenkel, M., Garti, N., and Kasan, R.,** Multiple emulsions. II. HLB shift caused by emulsifier migration to external interface, *J. Colloid Interface Sci.,* 97, 374, 1984.
31. **Chilamkurti, R. N. and Rhodes, C. T.,** Transport across liquid membranes: effect of molecular structure, *J. Appl. Biochem.,* 2, 17, 1980.
32. **Brodin, A. F., Kavaliunas, D. R., and Frank, S. G.,** Prolonged drug release from multiple emulsions, *Acta Pharm. Suec.,* 15, 1, 1978.
33. **Matsumoto, S. and Kohda, M.,** The viscosity of w/o/w emulsions: an attempt to estimate the water permeation coefficient of the oil layer from the viscosity changes in diluted systems on aging under osmotic pressure gradients, *J. Colloid Interface Sci.,* 73, 13, 1980.
34. **Kita, Y., Matsumoto, S., and Yonezawa, D.,** Viscometric method for estimating the stability of w/o/w-type multiple phase emulsions, *J. Colloid Interface Sci.,* 62, 87, 1977.
35. **Matsumoto, S., Inoue, T., Kohda, M., and Ohta, T.,** An attempt to estimate stability of the oil layer in w/o/w emulsions by means of viscometry, *J. Colloid Interface Sci.,* 77, 564, 1980.
36. **Tomita, M., Abe, Y., and Kondo, T.,** Viscosity change after dilution with solutions of water-oil-water emulsions and solute permeability through the oil layer, *J. Pharm. Sci.,* 71, 332, 1982.
37. **Matsumoto, S. and Sherman, P.,** A preliminary study of w/o/w emulsions with a view to possible food applications, *J. Texture Stud.,* 12, 243, 1981.
38. **Hashida, M., Yoshioka, T., Muranishi, S., and Sezaki, H.,** Dosage form characteristics of microsphere-in-oil emulsions. I. Stability and drug release, *Chem. Pharm. Bull.,* 28, 1009, 1980.
39. **Pilman, E., Larsson, K., and Tornberg, E.,** Inverse micellar phases in ternary systems of polar lipids/fat/water and protein emulsification of such phases to w/o/w-microemulsion-emulsions, *J. Dispersion Sci. Technol.,* 1, 267, 1980.
40. **Burbage, A. S. and Davis, S. S.,** The characterization of multiple (w/o/w) emulsions using a radiotracer technique, *J. Pharm. Pharmacol.,* 31, 6P, 1979.
41. **Brodin, A. F. and Frank, S. G.,** Drug release from o/w/o multiple emulsion systems, *Acta Pharm. Suec.,* 15, 111, 1978.
42. **Chiang, C.-W., Fuller, G. C., Frankenfeld, J. W., and Rhodes, C. T.,** Potential of liquid membranes for drug overdose treatment: in vitro studies, *J. Pharm. Sci.,* 67, 63, 1978.
43. **Yang, T. T. and Rhodes, C. T.,** Transport across liquid membranes: effect of formulation variables, *J. Appl. Biochem.,* 2, 7, 1980.
44. **Magdassi, S. and Garti, N.,** Release of electrolytes in multiple emulsions: coalescence and breakdown or diffusion through oil phase?, *Colloids Surf.,* 12, 367, 1984.
45. **Davis, S. S.,** The emulsion — obsolete dosage form or novel drug delivery system and therapeutic agent?, *J. Clin. Pharmacol.,* 1, 11, 1976.
46. **Stroeve, P. and Varanasi, P. P.,** Transport processes in liquid membranes: double emulsion separation systems, *Sep. Purif. Methods.,* 11, 29, 1982.

Chapter 3

GELS FOR DRUG DELIVERY

David W. Woodford and Dean S. T. Hsieh

TABLE OF CONTENTS

I. INTRODUCTION

The primary focus of this chapter is on the utility of naturally occurring gel-forming materials and their semisynthetic analogs. Historically, gels are ubiquitous, with formulated uses inspired by natural examples, logical extensions, and serendipity. A few obvious natural gels are acacia, tragacanth, pectin (in apples, oranges, strawberries, and lemons), slimy algal mucopolysaccharides and the bacterial mucopolysaccharides which allow adhesion to aquatic plants and dental enamel, bone and hide gelatin, and vitreous humor of the eye.

A common feature of such gels is a physical barrier property. It is responsible for reduced rates of evaporative water loss and retarded distribution of liquefactive enzymes, exotoxins, and whole organisms or viruses in the case of gelatinous defensive exudates of a prey or host organism. Chemically, the dispersed phase materials contained in these gels fall into the categories of polysaccharides and polyamides.

These systems were first identified by the term *colloid,* derived from *kolla,* the Greek word for glue. Thomas Grahm coined this term in the first chemistry paper on colloids. He defined colloids as certain aqueous dispersions of polypeptides such as gelatin and albumin, acacia, dextrin, and starch vegetable gums. Other colloids were inorganic, e.g., gelled metal hydroxides and Prussian blue. Crystalloids were cited as a distinct class. Unlike the colloids, salt and sugar could be induced to crystallize out of solutions and could be separated from mixtures containing colloids by dialysis. Crystalloids diffused through pores in animal gut membrane, whereas colloids were retained.[1] There is also evidence of ancient use of gels as glues.[2] References to various glues are found in the classical literature of Lucretius, Pliny, Shakespeare, and Marlowe.[3]

II. DEFINITION OF GELS

Gels are defined in the USP as semisolids consisting of inorganic particles (large or small) which are both enclosed and penetrated by a liquid. In addition to the familiar hydrogels mentioned above, organogels fit this definition. One example is mineral oil combined with polyethylene resin, which may form a gelatinous ointment base.[4] In classical colloid terminology, a gel is defined as a system having a characteristic cross-linked network of polymer chains which forms at the gel point.[5] Two principal types of gel systems employed for pharmaceuticals are those in the gel state at the time of application and those applied or administered in the dry state as a bolus, a film, or discrete particles. Examples of gel systems belonging to these two categories are shown in Table 1.

Gels may also be classified into two groups on the basis of porosity: (1) homogeneous gels, which have comparatively minute pores and undergo syneresis upon solvent loss, and (2) permanently porous gels, which have large pores and do not show network collapse upon loss of solvent; they may be macroreticular or macroporous. However, exceptions to this classification system are known. Agarose gels, for instance, possess large pores and are rigid, but collapse upon loss of water.

III. PHYSICAL CHARACTERISTICS OF GELS

When force is applied to a body, the distortion in shape or size which results is called a deformation. Flow is a continuous change in degree of deformation over a period of time.[7] Flow is typical of a substance in a liquid state. However, liquids possess merely viscosity, whereas a gel exhibits the following definitional properties:

1.   A yield value that is evident without an arbitrary, brief time interval
2.   Elasticity
3.   Viscosity

Table 1
PRINCIPAL TYPES OF GEL SYSTEMS

Gels formed *a priori*	Gels formed *in situ*
Aluminum hydroxide antacid gel	Capsule containing powdered hydroxypropyl methylcellulose and/or hydroxyethyl cellulose plus active drug
Flavored gelatin containing theophylline, extemporaneously prepared	Compressed or molded tablet containing the above cellulose excipients plus drug
Polyvinyl alcohol transdermal gel disk containing nitroglycerin	Cellulosic coatings on beads, granules, and tablets
Aloe vera mucilage	Coatings consisting of other hydrophilic gums or resins
Psyllium seed extract bulk-forming laxative	

Table 2
COMMON EXAMPLES OF VISCOELASTIC MATERIALS

Example	Classification	Observable properties
Molasses, 4°C	Liquid	Viscosity, yield value
Taffy, 25°C	Liquid	Viscosity, yield value
Glass, 25°C	Supercooled liquid	Yield value, elasticity
Glass, 1000°C	Liquid	Viscosity
Grape	Brittle gel	Yield value, elasticity
Gelled gelatin	Gel	Yield value, elasticity
Pudding	Soft gel	Elasticity, viscosity

All three components may be present in a particular body, although one or more may be so minor as to escape observation.[8] For example, gelatin gels may possess both viscous and elastic properties at a 1% w/v concentration at 20°C.[9] The elasticity of certain gels may be due to the presence of a double helix structure, similar to a molecular spring.[10,11] A review of the measurement of the dynamical mechanical properties of food gels has been published. It includes an extensive evaluation of available test methods and detailed mathematical models of gel rheology.[12]

Molecular polymer physics has been addressed in terms of its role in producing the observed macroscopic rheological behavior of gels.[13] Structural bonding, internal binding forces, and gel formation in cellulose ether hydrogels and in colloidal suspensions have been studied. Enslin numbers, thixotropy, flow, and viscosity curves have provided quantitative values. Gel formation has been closely related to water uptake capacity and to the rheological profile for each polymer tested.[14]

A nongel has no elasticity on the macroscopic scale — only on the small scale by reason of surface tension, e.g., the elastic quivering of small dewdrops on a blade of grass. A gel behaves elastically if forcefully distorted by either static or intermittent stresses that produce strains with a magnitude that is below the elastic limit of the gel. Upon cessation of the application of force, the deformed gel returns to its original geometry in a completely reversible fashion. However, application of greater force results in a displacement or strain beyond the elastic limit. This will cause macroscopic rupture or fragmentation of the brittle type of gel or microscopic rupture and viscous flow or liquefaction of the soft type of gel. Familiar examples of the properties of gels are shown in Table 2.

Loss of solvent markedly changes the characteristics of gels. Hydrogels exposed either to air or to solid surfaces that are not saturated with water tend to dehydrate. This produces a drier, more dense outer skin or membrane. If the environment is very dry, the gel functions as a glue. It eventually will become very brittle and strong. Dried films or boluses of natural gums are quite brittle and frangible and may be rendered pliable and tough (similar to the consistency of taffy candy) by the addition of plasticizers such as 8 to 10 wt% glycerin, diethylene glycol, ethylene glycol, or sorbitol.[15] As in the case of hydrogels, organogels may lose solvent and harden to form a brittle coating. The solvated, rubbery state may be preserved by an impermeable protective covering, a sufficient environmental solvent concentration, or a nonvolatile solvent (plasticizer).

Other properties of gels include syneresis and swelling. In the process of syneresis, gels contract, i.e., squeeze out the solvent in a sponge-like fashion. In swelling, a gel will absorb solvent from the environment. An example of syneresis is the appearance of liquid water on top of yogurt-containing carrageenan thickeners.

Of the many natural gums, only a few display true gel formation. In low concentrations, acacia, agar, pectin, sodium alginate, tragacanth, and xanthan gum are variously soluble or dispersible in either cold or hot water and remain liquid. However, higher concentrations and altered conditions result in gel formation. Agar and gelatin are soluble in hot water. Solutions thus formed produce gels when cooled since these materials are insoluble in cold water. Agar, for instance, requires heating in near-boiling water, gelling at 50 to 60°C; this very brittle gel will not melt at less than 80 to 90°C. Pectin, also soluble in hot water, produces smooth gels with high solids content in the presence of sugar and acid. Raw starch is soluble in hot water, forming a gel upon cooling. In the presence of phosphates, precooked or pregelatinized starch forms a gel (e.g., in cool aqueous liquids such as milk). Alginates form brittle, aqueous gels in both hot and cold media in the presence of certain ions. Carrageenan and furcellaran produce thermally reversible gels if dissolved in hot water and subsequently cooled. Guar gum will form a gel in the presence of borax, although it is not suitable for internal use. Carboxymethylcellulose solutions gel in the presence of aluminum ions. Low methoxy pectin dissolves in cold water and may form brittle gels with the addition of certain salts.[16]

The property of adhesion has been utilized for manufacturing lozenges capable of sustaining the release of drugs when placed in the oral cavity. Drug release occurs only after saliva hydrates the lozenge exterior to form a superficial gel.[17] Hydration of dried gels or powdered gel-forming polymers is the basis for many controlled release tablets and capsules that are commercially available. Examples of such products and details of manufacturing and formulation considerations follow in subsequent sections of this chapter.

IV. CONTROL OF DRUG RELEASE RATE USING GELS

Many drugs dissolve rapidly in the GI fluids to form solutions. Most drugs intended for oral administration are in use partly because they are readily absorbed and exhibit reasonably high bioavailability, which is dependent on their ability to pass through gut membranes. In the absence of active transport, membrane permeation is greatly increased if some portion of the total drug present in the solution exists in nonionized form. The absorbing surface area of the gut is large. Also, the rate of blood flow in the gut wall is high; this maintains the drug concentration at the absorbing surface at a low level or sink condition. As predicted by Fick's law of diffusion, this results in rapid systemic absorption of the drug in solution in the gut. For certain drugs, this may result in high initial blood concentrations accompanied by undesirable or even toxic side effects in addition to the intended effect.

There are several major effects that are important to the control of drug release from dosage forms that contain gel-forming components. These are

1.   Drug dissolution (drug suspended in matrix)
2.   Diffusion of dissolved drug through the matrix
3.   Erosion (gel matrix dissolution)

4.   Swelling due to absorption of water by the gel matrix
5.   Gel matrix disintegration

The characteristics of hydrogels may be exploited for controlled release drug delivery in one of two ways: (1) to slow the rate of drug dissolution for the drug-suspension type of gel matrix and (2) to allow the rate of drug release from the matrix surface into the gut for the drug-solution type of gel matrix. In either case, the result is a slower rate of absorption of drug into systemic circulation.

A principal characteristic of gels is the interspersed liquid continuous phase. In a gel matrix, this liquid is immobilized or stagnant due to entrapment by the dispersed phase material. One analogy is that of a sponge soaked with water. If dissolved drug is added to the liquid, the sponge represents a controlled release gel matrix. The continuous phase regions contain both drug molecules existing in a pseudogaseous state and liquid molecules of the medium. To initiate drug release, the sponge may be placed in a vessel containing pure water. Drug molecules tend to move from regions of higher concentration to regions of lower concentration (the vessel of water in which the sponge is situated) in the process known as diffusion. At the same time, the liquid water will self-diffuse in the opposite direction (from the vessel into the drug matrix). For viscoelastic gels, however, this sponge analogy is imperfect. In reality, the dispersed, discontinuous phase in natural hydrogels consists of filaments on the molecular scale. These filaments or individual molecules are neither stationary nor rigidly interconnected. As a result of bombardment by the water molecules of the continuous phase, they are in constant agitation. This phenomenon is known as Brownian motion. In Brownian motion, polymer molecules exist in a diluted, rubbery (or gel) state. They have vibrational and some rotational motion. In the liquid state, there is also translational motion. Drug molecules diffusing out of the hydrogel matrix encounter no fixed barriers. Thus, the apparent diffusion rate is nearly the same in low volume-fraction hydrogels as in pure water, although the diffusion rate of the drug molecules is reduced as the volume fraction of the matrix occupied by polymer increases.[18]

Swelling of a gel matrix may occur for hydrogels that require a low shear environment for the gel state and are yield-value dependent, such as xanthan gum hydrogels. The increased water content results in decreased volume fraction of the polymer, which translates into decreased diffusivity of the drug. However, this may be offset by the increased distance of diffusion in the enlarged matrix. A secondary effect of swelling of dilute hydrogels may be a decrease in gel strength, which increases the rate of erosion of a shear-sensitive gel matrix. Dense gels or dried gels are extreme cases in which the polymer concentration is high enough to produce appreciable osmotic pressure. Such materials may swell to 20 times their initial volume in the process of cell formation.

Fragmentation of a sustained release gel matrix initiates disintegration. There is an increase in surface area for drug release and an attendant decrease in the average diffusional distance for drug molecules. The result is an accelerated drug release rate. Viscous flow of the outer surface of such shear-sensitive matrices causes the dispersed phase to mix with and dissolve into the liquid outside the matrix proper, so that the outer matrix layers liquefy. This process is commonly known as erosion. It causes rapid dissolution of solvated drug molecules or immobilized drug particles that were previously trapped within the matrix substance. Erosion also decreases the average diffusional distance for dissolved drug molecules with intact matrix cores. All drug molecules have a shortened pathway out of the gel material. Since the rate of drug diffusional movement is the same, the total amount of drug is released in a shorter period of time.

Erosion and disintegration of a gel may be prevented by cross-linking with polyvalent ions, e.g., calcium, in the case of sodium alginate gels. Covalent cross-linking agents may also be used, e.g., formaldehyde or glutaraldehyde gases are used for gelatin gels. A more

Table 3
TYPES OF DISPERSED-PHASE MATERIALS USED TO MANUFACTURE HYDROGELS

Agarose	Gum agar	Polyacrylamide
Alginates	Gum arabic	Poly(acrylic) acid and homologues
Alkyl and hydroxylalkylcellulose	Gum ghatti	Polyethylene glycol
Amylopectin	Gum karaya	Poly(ethylene) oxide
Arabinoglactan	Gum tragacanth	Poly(hydroxylalkyl methacrylate)
Carboxymethylcellulose	Hydroxyethyl cellulose	Polyvinyl alcohol
Carrageenan	Hydroxypropylcellulose	Polyvinylpyrrolidone
Eucheuma	Hypnea	Propylene glycol alginate
Fucoidan	Keratin	Starch and modified analogs
Furcellaran	Laminaran	Tamarind gum
Gelatin	Locust bean gum	*n*-Vinyl lactam polysaccharides
Guar gum	Pectin	Xanthan gum

gradual release of formaldehyde may be obtained by gentle warming of incorporated hexamethylene tetramine. The rate of drug dissolution from the surface of gel-entrapped solid drug particles or crystals is affected by the drug concentration at the matrix/solid interface of the particles of suspended drug. In the case of a slowly dissolving particle, dissolved drug may quickly diffuse away, causing a low drug concentration at the particle surface, with the result that release of drug is dissolution-rate limited. On the other hand, when drug particles dissolve rapidly, the slow step (i.e., the rate-limiting step in a serial process) is diffusion of dissolved drug molecules through the gel and out of the matrix.

V. MECHANISMS OF GEL FORMATION IN SUSTAINED RELEASE PRODUCTS

Hydrogels may be formed using a number of water-soluble or water-insoluble dispersible gums or resins such as those shown in Table 3. The edible, nontoxic examples given have three advantages: (1) they are readily available in large quantities from multiple suppliers, (2) they are derived from natural sources at low cost, and (3) they require minimal processing before utilization. Polymers derived from natural sources such as gelatin, albumin, dextrin, acacia, and tragacanth are metabolized after enteral or parenteral administration.

The cellulosic and synthetic examples such as methylcellulose and polyvinylpyrrolidone are not biotransformed, but are excreted intact and are normally immune to biological degradation except for gut symbiotes in ruminants and termites in the case of cellulosics. Methylcellulose and polyvinylpyrrolidone are also highly resistant to chemical and physical degradation under the conditions encountered in pharmaceutical processing or in the GI tract. Some are also nontoxic and edible. In the context of human and veterinary applications, certain organogels have also proven useful for those materials which are nontoxic and nonirritating.

Some newly available gel-forming products are of interest due to their unusual properties. Poly-Levulan®, a gum arabic replacement, is a microbial polysaccharide. Locust Bean Gum TIC Pretested CWS Powder® is a cold-water-soluble form of normally hot-water-soluble material. A refrigerated nitroglycerin container known as Tabpak® contains a special coolant gel that rapidly cools to $-20°C$ in 1 hr, yet remains under 25°C for 4 hr due to slow reheating properties.

Hydrogels are in the major class of lyophilic colloidal dispersions and in the specific subclass of hydrophilic colloidal dispersions. Hydrophilic colloidal dispersions consist of three major classes: (1) true solutions, (2) gelled solutions, and (3) particulated dispersions.

True solutions are formed by water-soluble polymers such as acacia and polyvinylpyr-

Table 4
FACTORS TRANSMUTING SOLS INTO GELS

Method	Factor	Example	Mechanism
Mechanical agitation	Unit frequency (pulse)	Application of static pressure to sol	Dilatant sand or bentonite slurry thickens and becomes solid
	Low frequency	Shaking	Plug flow in center of container
	Ultrasonic frequency	Operation of cleaning bath or parallel plate rheometer	Liquid sol becomes elastic gel
	Collisional energy transfer	Conductive heating or cooling	Thermal gelation of methylcellulose, Tetrox®, Polyox®
Electromagnetic radiation	IR	Radiative cooling or absorptive heating	Thermal gelation
	UV	Exposure to sunlight or xenon lamp	Aging of skin by cross-linking of collagen sol
	Gamma and X	Exposure to cathode ray tube or radioactive cobalt	Polymerization of ethylene
Changes in electrical potential	Operation of electrochemical cell		Precipitation of zwitterions at electrodes
	Addition of gaseous reagent	Exposure to formaldehyde or glutaraldehyde	Cross-linking of gelatin sols
	Addition of liquid or solid reagent	Macromolecular species — pectin or carrageenan	Coacervation of gelatin with pH change
	Low molecular weight species	Hydronium ion	Precipitation of sodium alginate from solution
	Other ions	Univalent	Salting out of methylcellulose by sodium chloride
		Polyvalent	Cross-linking of alginates by calcium, aluminum ions

rolidone. Gelled solutions, known as gels or jellies, are formed if the polymer concentration is high and/or a temperature is reached at which solubility is low. This occurs, for example, when starch or gelatin sets to a gel on cooling or when methylcellulose gels on heating. Particulate dispersions of discrete, minute fragments are exemplified by bentonite or by microcrystalline cellulose. The formation of a solid or semisolid gel from a fluid dispersion depends on factors such as those listed in Table 4. To be a stable, evenly distributed mixture of two or more components, the dispersion (such as a gum or resin) must be either soluble or dispersible in the continuous phase. It may contain cosolvents or surfactant dispersion aids.[20]

Cellulose methyl ethers exhibit temperature-dependent gelation. When the temperature of an aqueous cellulose ether solution is raised, it becomes hazy and more viscous. Above a threshold concentration, the solution will be readily transformed into a soft or firm gel. Below the threshold of 0.5%, a fluid, interrupted gel is formed. It is composed of discrete, minute gel fragments and water. Higher cellulose concentrations produce a lower thermal gelation point. Methocel® type E exhibits a 10°C lower gel point with a 2% increase in concentration, whereas Methocel® type F shows only a 4°C lower gel point. Generally, higher viscosity correlates with a lower gel point. Additives that form ions, such as salts, decrease the gel point, and more highly charged ions are more effective. Some nonionic additives, such as sorbitol and glycerin, act similarly, whereas others raise the gel point.[21]

The various cellulose ethers in powdered form produce a cohesive, slowly dispersible gel when placed in water. This property has been utilized for prolongation of release of drugs contained in the cellulose ether gel. Slow erosion and slow disintegration or nondisintegration are the primary considerations in designing gel-forming matrices. Adding a soluble filler such as lactose increases the release rate of drug from the matrix. For rapidly dissolving drugs and excipients, only the K-type of Methocel® (hydroxypropyl methylcellulose, HPMC 2208 USP), which hydrates most rapidly, is effective in slowing drug release. This is because the superficial gel layer forms with sufficient rapidity to prevent disintegration of the dosage form when the other rapidly soluble matrix ingredients dissolve.

Tablets or capsules providing sustained release of drug may be formulated using approximately 10% HPMC. Higher concentrations of HPMC increase the sustaining effect on drug release. The release rate of matrices containing Methocel® type K15M may be further decreased by partial or total substitution with a newer grade of HPMC, Methocel®, in the same concentration. Other types of cellulose ethers in order of descending hydration rate are HPMC USP 2910 (Methocel® type E), HPMC USP 2906 (Methocel® type F), and Methylcellulose USP (Methocel® type A).[22] If the drug or other ingredients in the dosage form are slow to dissolve, any of the above types of Methocel® will sustain drug release, as is the case with pregranulated vitamin C (Roche C-90®). For slowly soluble ingredients, the erosion rate of the cellulose ether gel surface must be slower than the erosion rate of the other ingredients. More slowly hydrating types of cellulose ethers produce a more prolonged release of drug. For matrices containing low levels of insoluble fillers or poorly soluble drugs, the available surface area for wetting is decreased. Water permeability and subsequent drug release rate may also be decreased. Higher amounts of insoluble filler may produce faster drug release due to surface cracking caused by nonuniform swelling of the gelled cellulose ether component. For example, 7.5% of Methocel® type A15 produces slower release of vitamin C than does 20% Methocel® type A15. Insoluble excipients, either drug or additive, may be swellable or nonswellable in water. A swellable and insoluble additive such as microcrystalline cellulose (Avicel®) or crosscarmellose USP (Ac-Di-Sol®) may enhance the initial burst phase of drug release, shortening the 30% release time, with only a minute effect on the 90% release time. However, an insoluble and nonswellable additive such as dicalcium phosphate (Di-Tab®) may produce cracks in the matrix due to nonuniform swelling. Soluble salts compete for water availability for hydration of the cellulose ethers, slowing hydration and superficial gel formation resulting in a faster drug release rate.

A large proportion of small particles in the cellulose ether content assures a sufficiently rapid hydration rate and suitable prolonged drug release properties. A desirable specification for HPMC, Methocel® type K4M is 40% retained on a 200-mesh sieve screen and passing 100-mesh screen, with 30% passing a 200-mesh screen. Processing by wet granulation to prehydrate the cellulose ethers or by direct compression produces the same release rates, provided that the gelling agent is uniformly distributed throughout the matrix.

VI. METHODS OF GEL PREPARATION

A new procedure for making gels from sodium carboxymethylcellulose has been devised by workers in the Hercules Corporation. Aluminum formoacetate and the use of chromium salts and aluminum salts cause solutions of sodium carboxymethylcellulose to form gels which have high stability and a variety of textures.[24] Solid, transparent jellies may be manufactured by cooling aqueous solutions of $CaCl_2$, $Ba(SCN)_2$, or $Al_2(SO_4)_3$ or by cooling hydroalcoholic solutions of NaCl, KCl, NH_4, KSCN, NaBr, or NH_4NO_3, using solutions which are nearly saturated at room temperature. Gels are formed when aqueous solutions of carboxylated polymers are shocked by exposure to liquids of pH below 3.0. Examples include solutions of sodium alginate, carbomer (Carbopol 934®), and sodium carboxymethylcellulose.

Thermal gels may be produced from solutions of methylcellulose, hydroxypropyl methylcellulose, and polyethylene oxide. These materials are soluble in cold water, but precipitate to form gels in hot water. Gelatin forms thermoreversible gels: a 10% solution gels at 25°C, a 20% solution at 30°C, and a 30% solution at 32°C. Higher gelation temperatures are observed at the isoelectric point, at which pH value there exists a maximal interchain coulombic attraction. Stability of the ionic complexes decreases in the following order: Ba^{2+}, Cd^{2+}, Sr^{2+}, Ni^{2+}, Ca^{2+}, Zn^{2+}, Co^{2+}, and Mn^{2+}. Cu^{2+} and Cd^{2+} also complex with singular hydroxyl groups as well as vicinal diols.[27] A thermoreversible, highly cohesive colloid gel is obtained by making aqueous mixtures of xanthan gum and locust beam gum. Gel strength is maximal between xanthan/locust bean gum ratios of 6:4 to 4:6 and between pH 6 through 8. The mixture is heated to 180°F and gels between 120 and 130°F. Reheating above this range liquefies the gel.[28] Stable low methoxy pectin gels may be formed using material containing a minimum of 50% of the galacturonic acid in the free carboxyl form. This yield is obtained by precipitation of acid and NaOH deesterified pectin below pH 3 or precipitation of acid and Na_4OH deesterified pectin below pH 1.5. The free carboxyl form of the galacturonic acid reacts with calcium ions and gels by formation of strong cross-linkages between pectin chains. A total of 50 to 60 g of water in gel form can be held by 1 g of low methoxy pectin.[29] Casein and dialdehyde starch in a ratio of 10:1 form a gel in which two polypeptide chains are linked by a single molecule of dialdehyde starch.[30]

A sustained action topical base known as Plastibase® or Jelene® is available which is composed of 5% low molecular weight polyethylene and 95% mineral oil. The polyethylene is dissolved in the mineral oil at over 90°C. Subsequent chilling precipitates out the polymer to form a gel. Upon reheating to 60°C, the gel remains intact.[1]

Gels may be produced from polymer solutions that are nearly saturated with polymer by additions of inorganic salts. These salts bind part of the water, resulting in dehydration of the polymer, precipitation, and finally gelation. Ammonium sulfate is an example of an effective salt. This type is commonly termed a simple coacervate and is not a solid gel, as the addition of water liquefies the gel.

Aqueous polymer solutions may be induced to gel with the addition of alcohol. As a nonsolvent for the polymer, alcohols lower the dielectric constant of the liquid phase, lowering its ability to solvate the polymer. Also, the alcohol associates with water and competitively dehydrates the hydrophilic solvated polymer.[1]

The relationship between the molecular structure of gels and their macroscopic behavior has been investigated by sophisticated physicochemical methods. Computer modeling of pectin chain conformation and pectin gel formation has been accomplished and pectin chemistry has been reviewed.[31] Polygalacturonic acid gels have been investigated by proton and ^{13}C NMR. Gel formation is produced by multichain association beginning at 0.3 equivalents of calcium ion per carboxyl group. Gels may be formed by utilizing the principle of charge neutralization between two components in an aqueous solution: an anionic surfactant plus a cationic surfactant, an anionic polyelectrolyte and a cationic surfactant, or a cationic polyelectrolyte plus an anionic surfactant.

Aqueous dispersions of tragacanth form gels. Tragacanth is one third tragacanthin, which is water soluble, and two thirds bassorin, which swells in water to form a gel, but does not dissolve. Starch may also form aqueous gels. It is one sixth amylose, which is soluble in hot water, and five sixths amylopectin, which is insoluble by virtue of extensive branching of its chains. The amylopectin fraction absorbs water and swells to become a gel.[1] Aluminum hydroxide gel is produced by mixing 4% w/w of aluminum oxide with water. The resulting gel contains aluminum hydroxide and hydrated oxide. Aluminum phosphate gel is formed from a mixture of 4 to 5% $AlPO_4$ in water.[25] Silica gel is made by the reaction of sodium silicate with either hydrochloric acid or sulfuric acid, and the gel contains $(SiO_2) \cdot xH_2O$.[26] Alginic acid in aqueous solution is cross-linked by metal ions to form a strong gel. Experimentation has revealed that certain metals can induce cross-linking and establish the relative stability of the resulting complexes. Gelation occurs through cross-linkage of divalent metal ions with the carboxyl groups on a first macromolecule and with pairs of vicinal diols on a uronic acid residue on a second macromolecule. The calcium ion produces cross-linking of polymer chains to form a cooperative egg-carton structure that is characteristic of the rigid gel state.[32] Silicon dioxide gel formation from water glass solutions by ion exchange was studied using ^{29}Si NMR.[33] A very pure ferric oxide hydrate gel was prepared by hydrogen peroxide oxidation of a $FeC_2O_4 \cdot 2H_2O$ suspension, the result being a pure $Fe_2O_3 \cdot xH_2O$ gel. X-ray diffraction and electron microscopy revealed a very small crystalline character.[34] Gelatin which contains from 4 to 20 wt% of protein at 10°C was studied by high-resolution NMR. It was found that gelling began by nucleation of a collagen type of helix and then continued as a second-order reaction, analogous to the globule-to-helix transition in aqueous protein solutions. Bonding and reorientation of water molecules with decreased mobility of water molecules was found to occur during helix formation of phase gelation. Loss in water mobility paralleled increased protein content for gels.[36] Gelation of soybean globulin solutions has been studied as a function of soybean protein and sodium chloride concentration by means of differential adiabatic scanning calorimetry.[37] Milk gelation was described by a model evolved to characterize vulcanization of rubber. The modulus of final gels was proportional to the square of material concentration, and the gel-breaking strength was proportional to the modulus. The process was described by the general theory of gel formation by the cross-linking of polymer molecules.[38]

VII. PHARMACEUTICAL PATENTS

A number of patents have been granted pertaining to the use of gel-forming materials for controlled release of drugs. The following formulations exemplify systems based upon naturally occurring ingredients or their semisynthetic analogs.

Dry gelatin-coated drug spheres or pellets may be cross-linked by exposure to formaldehyde gas or solution. The process is controlled by dehydration in absolute alcohol using cupric sulfate and warm air for drying.[39] Subsequent immersion in water produces an insoluble hydrated cross-linked gelatin membrane that controls drug release. The cross-links are covalently bonded so that the resulting gel is noneroding.

Another system[40] utilizes Thixin-R®, a specially modified hydroxyl glycerol ester of an 18-carbon monobasic acid of acid value 2. It is further described as a modified 1-hydroxy stearin and a glycerol partial ester of stearic acid. When mixed with organic solvents such as methylene chloride or methanol, it forms a colloidal dispersion at room temperature. The substance is claimed to convert an organic solvent system into a semirigid mass, forming thread-like aggregates with three-dimensional structure. Isothermal, reversible sol-to-gel transformation and thixotropy have been observed. Incorporated drugs are released by combined erosion and diffusion mechanisms. Dried sols may be used for manufacture of compressed tablets. Inclusion of nonionic surfactants in the dried hydrophobic matrix allows water penetration in the test bath much like a wax type of matrix, in which water leaches drug out of the matrix by porous penetration.

A system containing casein and lipid has been described.[41] The fused lipid is mixed with casein and allowed to solidify and then made into granules which are compressed around an intact core of drug. Immersion in aqueous mixtures results in porous water penetration of the wax matrix variety together with hydration of the casein to form gelled channels. Drug release is by erosion, digestion, diffusion, and partitioning from the lipid into the aqueous medium.

A similar matrix system utilizes a dried coating or granulation containing a homogeneous mixture of drug and digestible fat or fat-soluble substance, either a gum or a proteinaceous material which swells in water to form gel channels.[42] The mixture may be formed by melting or solution. Faster drug release rates are claimed for glyceryl monostearate vs. stearic acid due to the more porous structure in the former. To obtain identical release profiles for the case of a water-soluble drug, a greater proportion of fatty additive is required, whereas a lipid-soluble drug requires a lesser amount of fatty material. This is due to partition control of drug release by fat in the latter drug. On the contrary, the proportion of swelling agent required is greater for the hydrophobic drug than the hydrophilic drug when identical release profiles are needed.

Lipid and silica were revealed as effective ingredients for controlled release.[43] A lipid of melting point 50 to 90°C (e.g., cetyl alcohol, stearyl alcohol, stearic acid, or palmitic acid) together with colloidal silica of less than 20 μm in particle size with a surface area of 100 to 400 m^2/g are used in a ratio of lipid to silica of 3:1 to 8:1. Silica, such as Cab-O-Sil®, is added to the melted lipid to form an optically clear gel with a viscosity directly proportional to the silica content. Drug of less than 60 mesh in particle size is added to the molten gel and the gel is cooled to form a solid. Next, the solid matrix is shaped into a bolus of the desired size or is broken into granules. Drug release is uniform and may be regulated in part by porous penetration of silica channels and subsequent swelling of the gelled silica.

A formulation was revealed which contains an earth or alkaline earth salt of a fatty acid (e.g., calcium, magnesium, or aluminum stearate) and polyvinylpyrrolidone (PVP) of k value between 24 and 40.[44] Its molecular weight varies from 40,000 to 80,000. It also contains a drug which is gradually released at time intervals controllable by the PVP/fatty acid salt ratio. Drug is dissolved in an aqueous PVP dispersion containing fatty acid salt macroparticles and then dried at 100°F. A thin film encloses the fatty acid salt particles, which are compressed into a tablet containing hydrous dibasic calcium phosphate. Exposure to digestive fluids results in hydration of the interparticle PVP porous channels to form gelatinous passages through which the drug diffuses from the tablet.

Another invention describes a formulation for the sustained release of pharmaceuticals in which powdered methylcellulose, pectin, and sodium alginate are mixed and made into a granulation by use of 75% isopropanol in water and then are compressed into tablets.[45] The methylcellulose absorbs water to form a gel, and the pectin and sodium alginate help to maintain the gel integrity by forming a coacervate.

A floating capsule that remains in the stomach for several hours before emptying into the intestine by means of a hydrodynamically balanced mixture of powders has been patented.[46] The formula contains hydroxypropyl methylcellulose. As the dried powder is contacted by gastric fluids, it is rapidly wetted and begins to hydrate. A tacky outer shell encloses the inner core of dry powder, and the capsule shape is preserved from disintegration. Buoyancy of the capsule (which contains air trapped within the mass of dry powder and eventually within the gel layer) causes it to be retained for several hours in the stomach, the preferred site of release for antacids and certain other drugs.

A system comprised of a tablet made of hydroxypropyl methylcellulose and nitroglycerin has been documented.[47] The tablet is placed in the buccal pouch, whereupon it absorbs water to form a gel which adheres to the buccal mucosa. It releases drug for a period of 6 hr.

Another innovative drug delivery system is composed of a dried hydrogel reservoir holding micropellets, each of which is simply a coated drug core.[48] The initially dry matrix tablet hydrates in the gastric fluids, expanding to 2 to 50 times its original volume. This expansion delays gastric emptying, thus prolonging the residence time of the bolus in the absorption window of the upper stomach and bowel.

A carrier base material consisting of hydroxypropylcellulose and up to 30% methylcellulose, sodium carboxymethylcellulose, and/or other cellulose ether has been revealed.[49] The base material is compressed into a solid dosage form. After ingestion, the matrix forms a gel that remains intact or slowly dissolves, so that a regular and prolonged release of drug results. Sustained release protein-bound 5-fluorouracil has been devised using a process of denaturing a mixture of 5% drug and 5% bovine albumin by exposure to 80°C for 5 sec.[50]

A soluble medicated vaginal insert composition has been disclosed which contains hydroxypropylcellulose and a drug such as miconazole nitrate, dienestrol, or a sulfathiazole, sulfacetamide, and sulfabenzamide combination. Purulent secretions are absorbed for improved hygiene during the course of infection. Moreover, secretions hydrate the polymer and cause drug to be released.[51]

An intravaginal collagen sponge has been developed which may contain Triton-X® 100. Native collagen is subjected to acid hydrolysis and the pH is adjusted to between 4.5 and 5.0 with acetic acid. The material is cross-linked by addition of glutaraldehyde, poured into a mold, and subjected to −10°C for 20 hr. Finally, a 2-hr wash in 20°C, pH 8 to 9 solution is followed by immersion in pH 4.5 to 5.0 buffer solution.[52]

It has been claimed that a certain collagen derivative is completely soluble and therefore useful in ophthalmic drug delivery because no residue remains after the drug is totally released. A 24-hr therapeutic effect was claimed for a pilocarpine-impregnated device. Calf skin digested by pH 2.5 HCl and pepsin for 5 days was filtered and neutralized to pH 10 with NaOH to inactivate the pepsin (4°C for 24 hr). Next, HCl was used to adjust the pH to between 7 and 8. The collagen precipitate was collected by centrifugation. After solubilization at pH 3 with HCl, the pH was raised to 9 using NaOH, and succinic anhydride in acetone solution was added. Acidification to pH 4.2 precipitated the succinylated collagen. This was dissolved at pH 3 and pilocarpine HCl was added. The pH was then raised to 7 using NaOH and vacuum deaeration was carried out. The mixture was poured into a mold and dried to form a membrane.[53]

A dry carrier composition for drugs produces prolonged release from lozenge, buccal, oral tablet, or suppository dosage forms. The carrier comprises from 40 to 70% of 50-cP hydroxypropyl methylcellulose, 20 to 40% of 4000-cP hydroxypropyl methylcellulose, and 5 to20% of hydroxypropylcellulose. Viscosities are for 2% aqueous solutions at 20°C. Low compression force produces compacts suitable for buccal, lozenge, or troche dosage forms, whereas high compression force produces harder compacts with a more prolonged release. Release rate of drug is controllable by varying the ratio of cellulose derivatives and by changing the compact surface area.[54]

Zein has been employed as a binder in formulations containing cetylpyridinium chloride, paradichlorobenzene, and salicylic acid. These zein hydrogels afforded sustained release of the bacteriostatic agents.[55] Sodium carboxymethylcellulose and carboxymethylene tablets containing nicotinic acid, diethylpropion, or an undisclosed antihistamine have been described as providing drug release over at least 4 hr.[56] Buccal or swallowed tablets comprised of hydroxypropyl methylcellulose in combination with ethyl cellulose have been disclosed. These tablets release a variety of drugs over 1 to 8 hr.[57]

Hydroxyethyl cellulose may be employed in the manufacture of pharmaceutical tablets, lozenges, and films, Drug may be released by gel formation, diffusion, and erosion. Examples include xanthine derivatives;[58] nitroglycerin, aminophylline, or potassium chloride (when combined with cetyl alcohol);[59] oral or nasal mucosal-adhesive delivery systems containing insulin or indomethacin;[60] theophylline or verapamil (combined with Eudragit®);[61] numerous drugs which are present as a mixture of free base and salt for the purpose of release rate adjustment;[62] indomethacin or niomethacin (when combined with waxes or ethyl cellulose or when granulated with organoaqueous solutions of acid-insoluble excipients such as carboxymethylcellulose, cellulose acetyl succinate, cellulose phthalate, or cellulose acetyl phthalate);[63] or benzodiazepines or digoxin (combined with suitable plasticizers, in the form of an edible multilayer film with drug only in interior layers, currently known as the Web Delivery System®).[64] Numerous additional patents depend on gel-forming ingredients. However, either one or more of these are synthetic materials. Thus, they are beyond the scope of this chapter. Comprehensive reviews of such patents may be found in other texts.[65,66]

VIII. EXPERIMENTAL GEL-BASED DRUG DELIVERY SYSTEMS IN THE LITERATURE

As demonstrated in Table 5, gel formation is the mechanism for prolongation of drug release for a great number of commercial products, both ethical and nonprescription. However, many other gel-forming sustained release systems are described in the literature. Some examples of these are given below.

- Pilocarpine has been incorporated into gels and gel-forming systems in order to prolong the duration of activity by controlled delivery rate. For example, a hydrophilic, thermoplastic film containing fibrin has produced an 8-hr duration of action.[67]
- Partial absorption of insulin has been demonstrated for oral mucosal formulations.[68]
- Granules made from sodium alginate may contain chemical herbicides.[69] Beads in the gel state were dried using a graded water/ethanol series and then critical-point dried in liquid carbon dioxide. These granules of dried gel absorb soil moisture which plasticizes the beads. This allows the herbicide to diffuse out of the rubbery matrix. However, in the glassy state, the dried gel beads do not allow the chemical to leach out.
- A soluble matrix gel for administration of pilocarpine over 24 hr after a single evening dose has been described.[70]
- Hydrocortisone acetate, a less water-soluble drug, has been dispersed in a hydrophilic, water-soluble matrix composed of partially cross-linked polypeptide.[71] Since release due to drug diffusion is limited by a slow dissolution rate for the drug, total drug release is governed by the dissolution or erosion of the matrix.
- The herbicide EPTC® may be incorporated into long-acting granules.[72] The herbicide is mixed into a neutral paste of flour or pregelatinized starch and then ammonia or an amine is added. Next, boric acid or a borate salt is used to cause gelation of the paste. Throughout the process, the pH is in the 9 to 11 range, but the final pH is between 7.5 and 8.2; this is an advantage with applications in delivery of pH-sensitive drugs.
- A hydrophilic matrix reservoir has been described which consists of an open-cell molecular sponge.[73] This structure contains cross-linked large molecular weight polysaccharides suspending a hydric alcohol and/or water solvent. Drug in solution or in suspension occupies the entrapped solvent volume that is immobilized by the polymer chains and cross-links. A commercial system using this technology is called Poroplastic®.
- Quinidine sulfate compressed tablets containing alginate, carrageenate, guar, or carob gum were found to sustain release of the drug.[74] A detailed investigation into the release mechanism was also conducted, and it was found that diffusion of water into the tablet and diffusion of drug out of or through the gelled layer were rate limiting.
- Pseudo zero-order release kinetics was reported for hydrogel membranes formed by polysalt bonding between oppositely charged polyelectrolyte polymers.[75] These complexes were used to coat microcapsules containing hemoglobin, insulin, and myoglobin. Zero-order release was produced which depended on a constant concentration of drug in the microcapsule core and a precise pore size for each drug entity.
- Pesticides may be dispersed in starch xanthan granules in order to provide a residual effect for compounds that would otherwise be washed away by rainfall or quickly volatilized.[76]
- An innovative hydrophilic laminate system is made of ethyl cellulose and hydroxypropyl methylcellulose.[77] It has a number of thin sheets of polymer, and inner layers contain more drug than outer layers. The laminated slab, much like a thin sheaf of paper, is rolled into a cylinder and placed in a gelatin capsule. As drug is depleted from the outer layer, the remaining drug molecules must diffuse a greater distance to

Table 5

EXAMPLES OF COMMERCIAL CONTROLLED RELEASE PRODUCTS THAT ARE DEPENDENT ON GEL FORMATION FOR FUNCTIONAL EFFICACY

Product or system trade name	Company	Description
Web Delivery System®	Hoffmann-La Roche	Cellulosic, laminated from multiple webs; 3 drugs, 1 water insoluble, in development
HBS®	Hoffmann-La Roche	Hydroxypropyl methylcellulose matrix capsule
Valrelease® Gastric Capsules		Diazepam
Gel-Kam® dental gel	R. P. Scherer Laboratories	Stannous fluoride, patented gel system
Gaviscon® Foamtabs	Marion Laboratories, Inc.	Antacid, alginic acid, and sodium bicarbonate
Gaviscon® Liquid	Marion Laboratories, Inc.,	Antacid, sodium acetate, and xanthan gum
Theodur Sprinkle®	Key Pharmaceuticals, Inc.	Theophylline, undisclosed propriety polymer mixture
Artificial Tears®	Ciba-Geigy Corp.	Tear substitute, methylcellulose
Synchron® System	Forest Laboratories	Hydroxypropyl methylcellulose
Not established	3M Laboratories	Theophylline, under development
Not established	3M Laboratories	Quinidine gluconate, under development
Site-Release® System	KV Pharmaceutical Co.	Undisclosed formula, highly bioadherent, liquid or semiliquid material, occlusive
Not established	Not revealed	Topical hydrocortisone
Not established	French firm, not revealed	Oral, liquid antacid
Not established	U.S. firm, not revealed, new drug application filed	Anticancer compound
Not established	Not revealed	Undisclosed formula, Nystatin, topical, under development
Not established	Not revealed	Topical benzocaine
Not established	Syntex Laboratories	Fluocinolone acetonide, topical, under development
Not established	Syntex Laboratories	Butoconazol nitrate, vaginal, under development
Not established	Stiefel Laboratories, Ltd.	Under development
Transderm-Nitro®	Key Pharmaceuticals, Inc.	Nitroglycerin, transdermal; polyvinyl alcohol, polyvinyl-2-pyrrolidone or agar, glycerol, sodium citrate, lactose, water
Not established	Merck, Sharpe and Dome Laboratories	Timolol, transdermal gelled mineral oil ointment, under development
Poroplastic®	Moleculon Biotech	Matrix/membrane system (cellulose triacetate with microporous structure)
Paratect® bolus	Pfizer device	Morantel ruminal depot
Not established	Heinrich Mack (Pfizer)	Isosorbide dinitrate transdermal device, under development
Not established	Thompson Medical (investigational new drug in preparation)	Phenylpropanolamine transdermal device
Not established	Thompson Medical	Triethanolamine salicylate transdermal device, under development
(Molecular sponge)	Pharmadyne (LecTec subsidiary)	Matrix delivery system, homogeneous microporous natural polysaccharide, cross-linked

Table 5 (continued)
EXAMPLES OF COMMERCIAL CONTROLLED RELEASE PRODUCTS THAT ARE DEPENDENT ON GEL FORMATION FOR FUNCTIONAL EFFICACY

Product or system trade name	Company	Description
Not established	Eli Lilly Co.	Insulin transdermal device, under development
Not established	Eli Lilly Co.	Nitroglycerin transdermal delivery device, under development

reach the matrix surface. The increased drug concentration in inner layers offsets the greater distance for diffusion, so that the drug release rate remains nearly constant. Surface layers are drug-free, semipermeable polymers and one side is more hydrophilic than the other, so that when the rolled laminate is wetted, it unfurls in a manner analogous to that of a bimetallic thermometer spring. Buoyancy is conferred by air bubbles trapped just below the outermost layers.

- Gelatin of the type used in pharmaceutical capsules and in food products may be cross-linked and denatured by ethylene oxide gas as used for sterilization. This may have an application in the formation of gel matrices.[78]

- Gelation of aqueous suspensions of cottonseed proteins has been reviewed with respect to a heat-stir process with varous conditions of pH, temperature, and concentration.[79]

- A gel was formed from a new polysaccharide produced by *Bacillus subtilis* FT-3 when aqueous suspensions of over 1% polysaccharide were heated in boiling water and allowed to stand at room temperature. The gel structure and main chain sequence were also reported.[80]

- Adhesive topical delivery systems have been prepared which use hydroxypropylcellulose and Carbopol® 934 for administration of insulin and bleomycin.[81]

- Biodegradable microparticles comprised of maltodextrin (cross-linked starch) and lysozyme, human serum albumin, carbonic anhydrase, or immunoglobulin G have been produced. Their in vitro release rates are between 30 and 100% after 12 weeks.[82]

- Microcapsules containing phenylpropanolamine have been produced which control drug release by an ethyl cellulose membrane coating.[83] An emulsion induction technique was employed.

- Hydrogenated ethanol-soluble soya lecithin has been found to produce opaque gels in concentrations over 8% in water or in ethanol. Drugs such as aminobenzoate esters, DL-α-tocopherol, and farnoquinone were admixed with nonhydrogenated soya lecithin in $CHCl_3$ to form clear films.[84]

- Ketoprofen, a poorly water-soluble drug, was incorporated into a number of gel-forming polymers: dextran T-40 and T-70, carboxymethylcellulose sodium, methylcellulose of three viscosity grades, sodium alginate, gelatin, carrageenan, pectin, gum arabic, gum tragacanth, guar gum, and locust bean gum. The drug and polymer were in each case dissolved using aqueous ammonium solution and then freeze-dried. Dissolution rates were determined using a rotating disk method. The dextrins caused the dissolution rate to increase by an order of magnitude, whereas tragacanth decreased the dissolution rate by an order of magnitude relative to pure ketoprofen.[85] A previous paper concerned dissolution of indomethacin from these gelling polymers.[86]

- A coprecipitate of pectin and phenothiazines produced initial dissolution rates from 100 to 2000 times slower than that of the pure drug for various phenothiazines. X-ray diffraction and infrared absorption studies were done to characterize the coprecipitate formed.[87]

- Gel-forming buccal tablets containing hydroxypropyl methylcellulose and ^{99m}Tc have been produced and the drug absorption rate was studied in human subjects by cephalic scintigraphy of the radioactivity remaining as a function of time. One subject showed 50% remaining after 3 hr in the lower buccal cavity vs. 50% after 1.5 hr in the upper buccal cavity and 50% after 1 hr sublingually.[88]

- Bovine serum albumin conjugates of methotrexate have been made and tested for physical characteristics, stability, in vitro release rate, and activity in tissue culture. Coupling was induced using a water-soluble carbodiimide in buffer. Dialysis testing showed under 10% free drug released after 24 hr.[89]

- A polyethylene-mineral oil gel, Plastibase 50W®, was used as a vehicle for salicylic acid suspensions. One suspension produced a 10% release of drug after 30 hr, and release was dissolution rate limited for a considerable time before diffusion rate-limited release predominated.[90]

- Sustained release oral lozenges have been developed utilizing the property of gel adhesion. Adhesion was found to be determined by polymer chain length and by the presence of ionizable pendant groups. Adhesion occurred after the polymer was hydrated by saliva in the oral cavity.[91]

- A general mathematical model for the controlled release of a drug from a right circular cylinder has been developed for the case of a porous matrix gel containing a homogeneous solution of the drug. Experimental results for the release of benzoic acid from 1% hydrogel cylinders closely matched the results predicted by the theoretical model used. The model accounted for the presence of a diffusion boundary layer at the matrix/release medium interface, degradation of drug inside the cylinder, variation in cylinder radius to height ratio, and internal porosity distribution of the matrix.[92]

- The effects of acacia and sodium alginate on the dissolution and disintegration of nicotinic acid tablets were investigated.[93] While 4% acacia had little effect, 4% sodium alginate markedly slowed both disintegration and dissolution of the drug.

- Methylcellulose, carboxymethylcellulose, and sodium alginate were reported to interact with platyphylline hydrotartrate in an equilibrium dialysis study.[94] The anionic polymers strongly inhibited release of the cationic form of the drug in solution, the result being prolonged action.

- Sodium alginate and propylene glycol alginate have been used in prolonged-action, intramuscular injectable formulations of epinephrine. The intent was to increase the vehicle viscosity.[95] Sodium alginate at a 2.5% level enhanced the disintegration rate of sulfanilamide from tablets, whereas levels of 20, 35, and 50% markedly retarded the release rate. The retarding effect was more pronounced in acid media compared to alkaline media.[96]

- Hydroxypropyl methylcellulose tablets containing potassium chloride were studied to determine the effect of entrapped air on the release rate. Higher compression force produced a release profile in accordance with the square root of the time, whereas lower compression force resulted in a linear release pattern with time. Evacuation of air from tablets also resulted in a square root-of-time release profile, regardless of the force used. High compression force was found to remove air, producing a less porous compact.[97]

- Water-insoluble glucan, derived from *Streptococcus mutans,* was used to produce a direct compression tablet containing chlorpheniramine maleate. Drug release was sustained, both in vitro and in tests with rabbits.[98]

- Hydroxyethyl methylcellulose tablets containing sodium salicylate were shown to release drug in accordance with an equation for diffusion from a two-sided slab matrix. During swelling in release medium, the tablets maintained a constant surface/volume ratio.[99] In a previous paper by the same authors, the effects of additives on the gellation temperature change, matrix integrity, and tablet release profile were reported.[100]

- In another study, tablets made from the following variety of polymers were manufactured: hydroxypropyl methylcellulose, methylcellulose, sodium carboxymethylcellulose, and two synthetics. Soluble drugs employed were chlorpheniramine maleate and sodium salicylate. Poorly soluble drugs used were benzoic acid and benzocaine. Gel formation, diffusivity of drugs in gels, and other parameters were determined.[101]

- A buoyant capsule containing hydroxypropyl methylcellulose and designed to remain in the stomach was studied by external gamma scintigraphy in human subjects. Gastric retention was observed for up to 6 hr. The delivery system is HBS®, Hydrodynamically Balanced System, and was previously detailed in Section VII.[102]
- A general perspective of the practical aspects of the use of cellulose ethers has been presented in a recent review paper.[103]

IX. CONCLUSION

It is evident that potential applications for gels in controlled release drug delivery are both numerous and varied. The unique capacity for gel delivery systems to limit the quantity of drug released while maintaining a constant dosage renders their future use unquestionably advantageous. Though clinical use is still far in the future for most experimental systems, the biocompatibility of most gels, the utilization of known drugs in present research, and the effective method of sustained drug delivery anticipate the practical use of gels as well as their market potential.

REFERENCES

1. **Osol, A., Ed.,** *Remington's Pharmaceutical Sciences,* 16th ed., Mack, Easton, Pa., 1980, 266.
2. **Bougue, R. H.,** *The Chemistry and Technology of Gelatin and Glue,* 1st ed., McGraw-Hill, New York, 1922.
3. **Smith, P. I.,** *Glue and Gelatine,* Chemical Publishing, Brooklyn, 1943, 2.
4. *The United States Pharmacopeia,* 20th rev., U.S. Pharmacopeial Convention, Rockville, Md., 1979, 1026.
5. **Hermans, P. H.,** Gels, in *Colloid Science,* Vol. 2, Kryut, H. R. Ed., Elsevier, Amsterdam, 1949, 483.
6. **Flodin, P. and Lagerkvist, P.,** Polymer gels, *J. Chromatogr.,* 215, 7, 1981.
7. **Fredrickson, A. G.,** *Principles and Applications of Rheology,* Prentice-Hall, Englewood Cliffs, N.J., 1964.
8. **Reiner, M.,** *Deformation, Strain and Flow,* Interscience, New York, 1960.
9. **Robinson, J., Kellaway, I., and Marriot, C.,** The effect of blending on the rheological properties of gelatin solutions and gels, *J. Pharm. Pharmacol.,* 27, 818, 1975.
10. **Rees, D. A.,** Polysaccharide gels: a molecular view, *Chem. Ind. (London),* p. 630, 1972.
11. **Rees, D. A.,** Double helix structure in foods, *Sci. J.,* p. 47, 1970.
12. **Rao, V. N. M.,** Dynamic force-deformation properties of foods, *Food Technol. (Chicago),* p. 103, 1984.
13. **Helfand, E.,** Dynamics of conformational transitions in polymers, *Science,* 226, 647, 1984.
14. **Regdon, G., Jr. and Eros, I.,** Rheological testing of cellulose ether hydrogels. I. Factors influencing the formation of the structure and the structural forces, *Acta Pharm. Hung.,* 54, 133, 1984.
15. **Braude, F.,** *Adhesives,* Chemical Publishing, Brooklyn, 1943, 68.
16. **Furia, T., Ed.,** *Handbook of Food Additives,* 2nd ed., CRC Press, Boca Raton, Fla., 1972, 301.
17. **Smart, J., Kellaway, I., and Worthington, H.,** An *in vitro* investigation of mucosa-adhesive materials for use in controlled drug delivery, *J. Pharm. Pharmacol.,* 36, 295, 1984.
18. **Lauffer, M.,** Theory of diffusion in gels, *Biophys. J.,* 1, 205, 1961.
19. **Braude, F.,** *Adhesives,* Chemical Publishing, Brooklyn, 1943, 76.
20. **Osol, A., Ed.,** *Remington's Pharmaceutical Sciences,* 16th ed., Mack, Easton, Pa., 1980, 267.
21. Methocel™ Product Information, Form #125-306-63, Dow Chemical Company, Midland, Mich., 1962, 24.
22. Formulating Sustained Release Pharmaceutical Products with Methocel™, Form #192-869-782, Dow Chemical Company, Midland, Mich., 1982, 9.
23. Methocel™ Product Information, Form #192-901-982, Dow Chemical Company, Midland, Mich., 1982.

24. **Anon.,** *Drug Cosmet. Ind.,* 134(2), 77, 1984.

25. *The United States Pharmacopeia,* 20th rev., U.S. Pharmacopeial Convention, Rockville, Md., 1979, 25.

26. *The United States Pharmacopeia,* 20th rev., U.S. Pharmacopeial Convention, Rockville, Md., 1979, 1253.

27. **Schweiger, R. G.,** Complexing of alginic acid with metal ions, *Kolloid Z. Z. Polym.,* 196, 47, 1964.

28. **Kovacs, P.,** Useful incompatibility of xanthan gum with galactomannans, *Food Technol.,* 27, 26, 1973.

29. **Padival, A., Ranganna, S., and Manjrekar, S.,** Mechanism of gel formation by low methoxy pectins, *J. Food Technol.,* 14, 277, 1979.

30. **Prahl, L., Schwenke, K., and Braudo, E.,** Studies on the mechanical properties of gels of chemically modified proteins, *Acta Aliment. Acad. Sci. Hung,* 5, 183, 1979.

31. **Nelson, D.,** Pectin — a review of selected advances made in the last 25 years, *Proc. Int. Soc. Citric.,* 3, 739, 1979.

32. **Rinaudo, M., Ravanat, G., and Vincedon, M.,** Nmr investigation on oligo- and poly(galacturonic acid)s; gel formation in the presence of calcium(2 +) counterions, *Makromol. Chem.,* 181, 1059, 1980.

33. **Grimmer, A., Starke, P., and Wieker, W.,** High resolution solid-state silicon-29 nmr of silica gels, *Z. Chem.,* 22, 44, 1982.

34. **Matzen, G. and Poix, P.,** Preparation of a very pure ferric oxide hydrate ($Fe_2O_2 \cdot xH_2O$) gel: some structural and magnetic properties, *J. Mater. Sci.,* 17, 701, 1982.

35. **Naryshkina, E., Volkov, V., Dolinnyi, A., and Izamaillova, V.,** Study of gelatin gel formation by high resolution nuclear magnetic resonance, *Kolloidn. Zh.,* 44, 356, 1982.

36. **Ksenofontov, B. and Kuleshov, I.,** Effect of physical actions on kinetics of formation of active silica gels used as a binding component, *Stroit. Mater.,* 3, 26, 1982.

37. **Bikbov, T., Grinberg, V., Danilenko, A., Chaika, T., Vaintraub, I., and Tolstoguzov, V.,** Studies on gelatin of soybean globulin solutions. III. Investigation into thermal denaturation of soybean globulin fraction by the method of differential adiabatic scanning calorimetry: interpretation of thermograms, the effect of protein concentration and sodium chloride, *Colloid Polym. Sci.,* 261, 346, 1983.

38. **Johnston, D.,** Application of polymer crosslinking theory to rennet-induced milk gels, *J. Dairy Res.,* 51, 91, 1984.

39. **Glassman, J. A.,** U.S. Patent 3,275,519, 1966.

40. **Tansey, R. P.,** U.S. Patent 3,136,695, 1964.

41. **Stephenson, D.,** U.S. Patent 3,184,386, 1965.

42. **Playfair, M. L.,** U.S. Patent 3,147,187, 1964.

43. **Lippman, I.,** U.S. Patent 3,400,197, 1968.

44. **Fennel, J. R.,** U.S. Patent 3,102,845, 1963.

45. **Klippel, K. R.,** U.S. Patent 3,558,768, 1971.

46. **Sheth, P. R. and Tossounian, J. L.,** U.S. Patent 4,126,672, 1978.

47. **Schor, J. M.,** U.S. Patent 4,226,849, 1980.

48. **Uruquhart, J. and Theeuwes, F.,** U.S. Patent 4,434,153, 1984.

49. **Schor, J. M., Nigalaye, A., and Gaylord, N. G.,** U.S. Patent 4,389,393, 1983.

50. **Kaetsu, I., Asano, M., Kumakura, M., and Yoshida, M.,** German Patent 3,104,815, 1982.

51. **Williams, B. L.,** U.S. Patent 4,317,447, 1982.

52. **Chvapil, M.,** U.S. Patent 4,274,410, 1981.

53. **Miyata, T., Rubin, A. L., Stenzel, K. H., and Dunn, M. W.,** U.S. Patent 4,164,559, 1979.

54. **Lowey, H.,** U.S. Patent 4,259,314, 1981.

55. **Cohen, A.,** U.S. Patent 4,007,258, 1977.

56. **Chriostenson, G. and Dale, L.,** U.S. Patent 3,065,143, 1962.

57. **Lowey, H.,** U.S. Patent 3,870,790, 1975.

58. **Gleixner, K.,** U.S. Patent 4,189,469, 1980.

59. **Leslie, S. T.,** U.S. Patent 3,965,256, 1976.

60. **Teijin, K. K.,** German Patent 2908-847, 1980.

61. **Deboeck, A. M. and Baudier, P.,** European Patent Application 142-877A2, 1985.

62. **Oshlack, B. and Leslie, S. T.,** European Patent Application 97-523A2, 1984.

63. **Hopfgartner, J., Hurka, W., Grablowitz, O., and Kropp, W.,** Belgian Patent 900-084, 1985.

64. **Hoffmann-LaRoche,** Belgian Patent 849-377, 1977.

65. **Colbert, J. C., Ed.,** *Controlled Action Drug Formulations,* Noyes Data Corporation, Park Ridge, N.J., 1974.

66. **Jones, D. A., Ed.,** *Transdermal and Related Drug Delivery Systems,* Noyes Data Corporation, Park Ridge, N.J., 1984.

67. **Miyazaki, S., Ishii, K., and Takada, M.,** Pharmaceutical applications of biomedical polymers. V. Use of fibrin film as a carrier for drug delivery: a long-acting delivery system for pilocarpine into the eye, *Chem. Pharm. Bull.,* 30, 3405, 1982.

68. **Ishida, M., Machida, Y., Nambu, N., and Nagai, T.,** New mucosal dosage form of insulin, *Chem. Pharm. Bull.,* 29, 810, 1981.

69. **Connick, W., Jr.,** Controlled release of the herbicides 2,4 D and dichlobenil from alginate gels, *J. Appl. Polym. Sci.,* 27, 3341, 1982.
70. **Mandell, A., Stewart, R., and Kass, M.,** *Invest. Ophthalmol. Vis. Sci. (Suppl.),* 165, April 1979.
71. **Dohlmann, C., Pavan-Langston, D., and Rose, J.,** A new ocular insert device for continuous constant-rate delivery of medication to the eye, *Ann. Ophthalmol.,* 4, 823, 1972.
72. **Sasha, B., Trimnell, D., and Otey, F.,** Starch-borate complexes for EPTC encapsulation, in 10th Int. Controlled Release Symp., Controlled Release Society, San Francisco, 1983, 311.
73. **Hymes, A.,** A solid state hydrophilic reservoir for transdermal delivery systems, in 10th Int. Controlled Release Symp., Controlled Release Society, San Francisco, 1983, 219.
74. **Bamba, M., Puisieux, F., Marty, J.-P., and Carstensen, J.,** Release mechanisms in gelforming sustained release preparations, *Int. J. Pharm.,* 2, 307, 1979.
75. **Lim, F.,** Polyelectrolyte complex microcapsules as controlled release carriers, in 10th Int. Controlled Release Symp., Controlled Release Society, San Francisco, 1983, 266.
76. **Shasha, B.,** Starch and other polyols as encapsulating matrices for pesticides, in *Controlled Release Technologies: Methods, Theory and Applications,* Vol. 2, Kydonieus, A., Ed., CRC Press, Boca Raton, Fla., 1980, 207.
77. Advances in sustained release of medications, *Chemical Week,* p. 56, May 11, 1983.
78. **Chaigneau, M. and Muraz, B.,** Action of ethylene oxide on gelatin: adsorption and reaction, *Sci. Tech. Pharm.,* 11, 269, 1982.
79. **Cherry, J. and Berardi, L.,** Heat-stir denaturation of cottonseed proteins: texturization and gelation, in *Food Protein Deterior. Mech. Funct. ACS Sym. Ser.,* 206, 1982, 163.
80. **Morita, N., Takagi, M., and Murao, S.,** A new gel-forming polysaccharide produced by *Bacillus subtilis FT-3.* Its structure and its physical and properties, *Bull. Univ. Osaka Prefect. Ser. B,* 31, 27, 1979.
81. **Merkle, H. P.,** Release kinetics of polymeric laminates for transdermal delivery, in 2nd Int. Recent Advances in Drug Delivery Systems, Controlled Release Society, Salt Lake City, 1985, 33.
82. **Artursson, P., Edman, P., Laasko, T., and Sjoholm, I.,** Characterization of polyacryl starch micro-particles as carriers for proteins and drugs, *J. Pharm. Sci.,* 73, 1507, 1984.
83. **Kaeser, L. B., Kissel, T., and Sucker, H.,** Manufacture of controlled release formulations by a new microencapsulation process, the emulsion-induction technique, *Acta Pharm. Technol.,* 30, 294, 1984.
84. **Matsumoto, M. and Fujii, M.,** Soya lecithin and related materials as pharmaceutical ingredients, *J. Pharmacobiodyn.,* 7, s-30, 1984.
85. **Takayama, K., Nambu, N., and Nagai, T.,** Factors affecting the dissolution of indomethacin dispersed in various water-soluble polymers, *Chem. Pharm. Bull.,* 30, 3013, 1982.
86. **Takayama, K., Nambu, N., and Nagai, T.,** Factors influencing the dissolution of ketoprofen from solid dispersions in various water-soluble polymers, *Chem. Pharm. Bull.,* 30, 673, 1982.
87. **Takashi, Y., Nambu, N., and Nagai, T.,** Interactions of phenothiazines with pectin in the solid state, *Chem. Pharm. Bull.,* 30, 2919, 1982.
88. **Hardy, J., Kennerly, J., Taylor, M., Wilson, C., and Davis, S.,** Release rates from sustained-release buccal tablets in man, *J. Pharm. Pharmacol.,* Suppl. 34, 1982.
89. **Halbert, G., Stuart, J., and Florence, A.,** Physical and biological characterization of methotrexate/albumin conjugates, *J. Pharm. Pharmacol.,* Suppl. 34, 1982.
90. **Graham, M., Proudfoot, S., and Ward, M.,** An in-vitro study of particle size on drug release from a polyethylene-mineral oil gel, Plastibase™ 50W, *J. Pharm. Pharmacol.,* Suppl. 34, 1982.
91. **Smart, J., Kellaway, I., and Worthington, H.,** An *in vitro* investigation of mucosa-adhesive materials for use in controlled drug delivery, *J. Pharm. Pharmacol.,* 36, 295, 1984.
92. **Tojo, K. and Chien, Y.,** Mathematical simulation of controlled drug release from cylindrical matrix devices, *Drug Dev. Ind. Pharm.,* 10, 753, 1984.
93. **Sakr, A. and Elsabbagh, H.,** Delayed release in compressed nicotinic acid tablets, *Manuf. Chem. Aerosol News,* p. 41, December 1973.
94. **Rachev, D., Dmitrievskii, D., and Pertsev, I.,** Interaction of platyphylline hydrotartrate with some polymeric substances, *Khim. Farm. Zh.,* 17, 1271, 1983.
95. **Ouer, R.,** New slow acting epinephrine solutions, *Ann. Allergy,* 11, 36, 1953.
96. **Salib, N. and El-Gamal, S.,** Application of some polymers in the physicochemical design of tablet formulation, *Pharmazie,* 31, 718, 1976.
97. **Korsmeyer, R., Gurny, R., Doelker, E., Buri, P., and Peppas, N.,** Mechanisms of potassium chloride release from compressed, hydrophillic, polymeric matrices: effect of trapped air, *J. Pharm. Sci.,* 72, 1189, 1983.
98. **Masumoto, K., Yoshida, A., Hayashi, S., Nambu, N., and Nagai, T.,** In vitro release profile and in vivo absorption study of sustained-release tablets containing chlorpheniramine maleate and water-insoluble glucan, *Chem. Pharm. Bull.,* 32, 3720, 1984.
99. **Touitou, E. and Donbrow, M.,** Drug release from non-disintegrating hydrophillic matrices: sodium salicylate as a model drug, *Int. J. Pharm.,* 11, 355, 1982.

100. **Touitou, E. and Donbrow, M.,** Influence of additives on (hydroxyethyl) methylcellulose properties: relation between gelation temperature change, compressed matrix integrity and drug release profile, *Int. J. Pharm.,* 11, 31, 1982.
101. **Lapidus, H. and Lordi, N.,** Drug release from compressed hydrophillic matrices, *J. Pharm. Sci.,* 57, 1292, 1968.
102. **Sheth, P. and Tossounian, J.,** The hydrodynamically balanced system (HBS™): a novel drug delivery system for oral use, *Drug Dev. Ind. Pharm.,* 10, 313, 1984.
103. **Alderman, D.,** A review of cellulose ethers in hydrophillic matrices for oral controlled-release dosage forms *Int. J. Pharm. Technol. Prod. Manuf.,* 5, 1, 1984.

Chapter 4

SYNTHETIC HYDROGELS FOR DRUG DELIVERY: PREPARATION, CHARACTERIZATION, AND RELEASE KINETICS

Ping I. Lee

TABLE OF CONTENTS

I. INTRODUCTION

Recent development in polymeric delivery systems for the controlled release of therapeutic agents has demonstrated that these systems can not only improve drug stability both in vitro and in vivo by protecting labile drugs from harmful conditions in the body, but also increase residence time at the application site and enhance the activity duration of short half-life drugs. Therefore, compounds which otherwise would have to be discarded due to stability and bioavailability problems may be rendered useful through a proper choice of polymeric delivery system. Furthermore, polymeric systems can provide predictable and reproducible drug release for extended duration in meeting a specific therapeutic requirement, thereby eliminating side effects, frequent dosing, and waste of drugs.

One such polymer system for drug delivery applications, which has attracted significant recent attention, is based on hydrogels. Hydrogels are hydrophilic network polymers which are glassy in the dehydrated state and swollen in the presence of water to form an elastic gel. Although hydrogels are of either natural or synthetic origin, it is the covalently cross-linked synthetic hydrogels that have been gaining increasing popularity in various biomedical applications, ranging from soft contact lenses to drug delivery systems,[1-5] ever since their introduction more than 20 years ago by Wichterle and Lim.[1]

Because of their relatively high water content and their soft, rubbery nature, several advantages can be readily realized for hydrogels. First, the high water content, permeability to small molecules, and cross-linked structure allow residual monomers and initiators to be efficiently extracted from the gel network before usage. Second, the soft and rubbery nature of most hydrogels can contribute to their biocompatibility as implants by minimizing mechanical irritation to surrounding tissue. In addition to the inertness and good biocompatibility of hydrogels, their ability to release entrapped drug in an aqueous medium and the ease of regulating such drug release by controlling water swelling and cross-linking density make hydrogels particularly suitable as drug carriers in controlled delivery applications.

Depending on the intended route of administration, drug-loaded hydrogel delivery systems are prepared into different geometries such as disks, rods, granules, microcapsules, and beads. Owing to their stability and dosing requirements, these drug-loaded hydrogel delivery systems are stored and administered either in the swollen, rubbery state for ophthalmic and implant applications or in the dry glassy state for oral delivery use. In the latter area, the most popular delivery system has been granules or beads where the drug is uniformly dissolved or dispersed in the hydrogel matrix, because of its lower cost and relative ease of fabrication. The release of water-soluble drugs from such dehydrated hydrogel matrices involves the simultaneous absorption of water and desorption of drug via a swelling-controlled diffusion mechanism. Such swelling-controlled diffusion generally does not follow a Fickian diffusion mechanism. The existence of some molecular relaxation process in addition to diffusion is believed to be responsible for the observed non-Fickian behavior.[5-7] Thus, hydrogels offer a unique combination of release mechanisms not readily available in other types of delivery systems.

In this chapter, the chemistry and methods of fabrication of hydrogel systems will be examined. Special emphasis will be placed on the suspension polymerization process for making hydrogel beads for drug delivery. In addition, the drug loading and release characteristics as well as methods of modifying release kinetics from hydrogel systems will be discussed. This chapter is not intended to be an exhaustive review of the field; rather, general concepts with specific examples will be given to illustrate various approaches in the design and fabrication of hydrogel drug delivery systems.

II. STRUCTURE OF SYNTHETIC HYDROGELS

In general, hydrogels for drug delivery applications are prepared by the polymerization or copolymerization of certain hydrophilic monomers with a cross-linking agent using free radical initiators.[8] Useful hydrogels are also obtained by copolymerizing and cross-linking of both hydrophilic and hydrophobic monomers to balance the desired level of equilibrium water absorption in the hydrogel. Examples of hydrophilic and hydrophobic monomers potentially useful for hydrogel preparation are listed in Table 1. Frequently used hydrophilic monomers include hydroxyalkyl esters of acrylic and methacrylic acids such as 2-hydroxy-ethyl, 2-hydroxypropyl acrylate or methacrylate, derivatives of acrylamide and methacry-lamide such as *N*-ethyl acrylamide and *N*-(2-hydroxypropyl) methacrylamide, and *N*-vinyl-2-pyrrolidone.[9] These monomers can be polymerized alone or in combination with each other and other suitable hydrophobic monomers. Commonly used hydrophobic monomers are alkyl acrylates or methacrylates such as methyl and ethyl methacrylate or acrylate, vinyl acetate, acrylonitrile, and styrene. Commonly used initiators include azo compounds, per-oxides, and redox catalyst such as azobisisobutyronitrile, *t*-butyl peroctoate, benzyl peroxide, ammonium persulfate, ammonium persulfate-sodium bisulfite combination, and diisopropyl percarbonate.

The cross-linking agent is usually a minor but important ingredient in synthetic hydrogels. In general, polymers that are not cross-linked are soluble in some solvents and they can be shaped or pressure-molded by heat. On the other hand, cross-linked polymers such as hydrogels will swell but not dissolve in solvents and they cannot be permanently formed by heat and pressure. Examples of difunctional cross-linking agents are also given in Table 1. The most commonly used cross-linking agents are ethylene glycol dimethacrylate (EGDMA), tetraethylene glycol dimethacrylate, and methylene-bis-acrylamide. A new class of cross-linking agent reported recently by Mueller and Good[10] is based on difunctional derivatives of hydrophobic polymers or oligomers such as poly-*n*-butylene oxide end-capped with is-ophorone diisocyanate. Such a high molecular weight cross-linking agent not only serves as a structural cross-link, but also imparts better mechanical strength to the swollen gel than those prepared with a short cross-linking agent.[11] Another promising development along these lines by Graham and McNeill[12] involves crystalline-rubbery polymers of poly(ethylene oxide) using diisocyanates and polyols to provide cross-linking.

III. PREPARATION OF HYDROGELS

Presently, the most commonly used hydrogels are poly(2-hydroxyethyl methacrylate) (PHEMA) and copolymers of 2-hydroxyethyl methacrylate (HEMA). Since the introduction of PHEMA for biomedical use in the early 1960s by Wichterle and Lim,[1] its principal commercial use has been in soft contact lenses.[3] PHEMA is perhaps also the most widely studied synthetic hydrogel. Several reviews on its properties and uses can be found in the literature.[2,3,8]

An important problem encountered in the preparation of PHEMA is the purity of monomers used.[13] Degradation of HEMA monomer during transportation and storage at ambient tem-peratures may result in increased level of impurities such as methacrylic acid, ethylene glycol dimethacrylate, and ethylene glycol according to the following reaction scheme:

$$CH_2=C(CH_3)COOCH_2CH_2OH + H_2O \rightarrow CH_2=C(CH_3)COOH + HOCH_2CH_2OH$$

$$2CH_2=C(CH_3)COOCH_2CH_2OH \rightarrow CH_2=C(CH_3)COOCH_2CH_2OOCC(CH_3)=CH_2$$

$$+ HOCH_2CH_2OH$$

The cross-linker EGDMA formed as an impurity by transesterification of two HEMA mol-ecules will result in the cross-linking of HEMA even when no cross-linker is added to the

Table 1
EXAMPLES OF MONOMERS USED IN HYDROGELS

Hydrophilic

Hydroxyalkyl acrylates or methacrylates

R = –H, –CH$_3$
R' = –H, –CH$_2$CH$_2$OH, –CH$_2$CH$_2$OCH$_2$CH$_2$OH,
–CH$_2$CHOH –CH$_2$CHCH$_2$OH –CH$_2$CH$_2$OCH$_3$
　　　　　　　　　｜
　　　　　　　　CH$_3$　　　OH
–CH$_2$CH$_2$OCH$_2$CH$_2$OCH$_3$

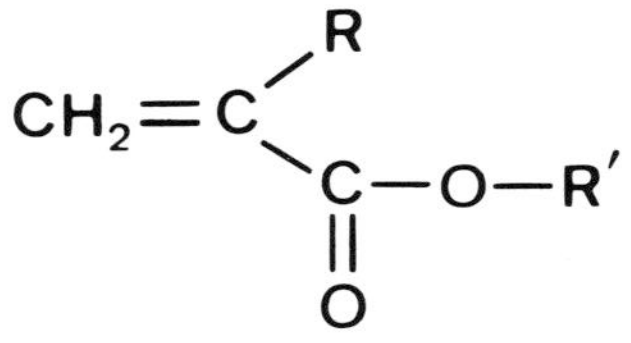

Acrylamide derivatives

R= –H, –CH$_3$
R', R'' = –H, –CH$_3$, –C$_2$H$_5$, –CH$_2$CHCH$_3$
　　　　　　　　　　　　　　　　　　　｜
　　　　　　　　　　　　　　　　　　OH

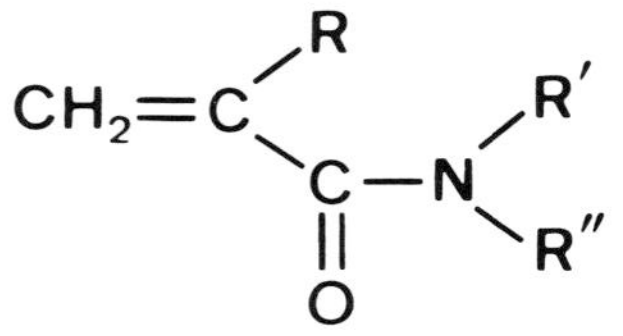

N-vinyl-2-pyrrolidone

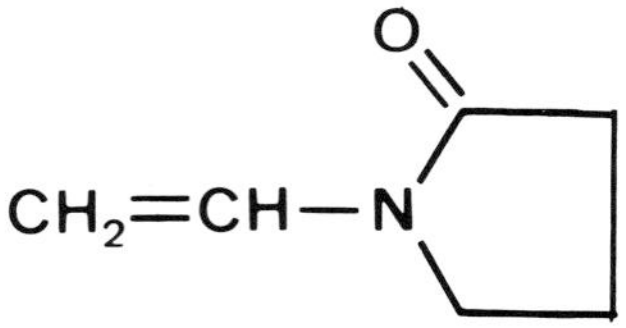

Hydrophobic

Acrylics

R = –H, –CH$_3$
R' = –CH$_3$, –C$_2$H$_5$, –C$_4$H$_9$, –OCH$_3$, –CN, –OCH$_2$CH$_2$OCH$_3$

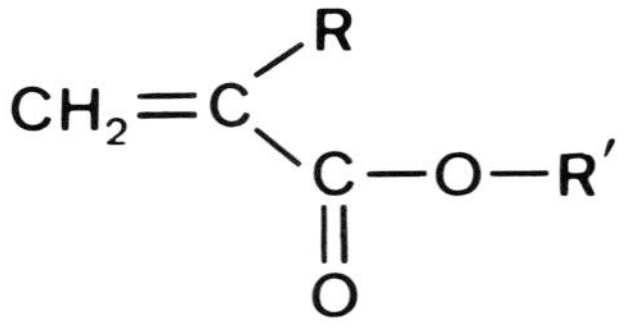

Vinyl acetate

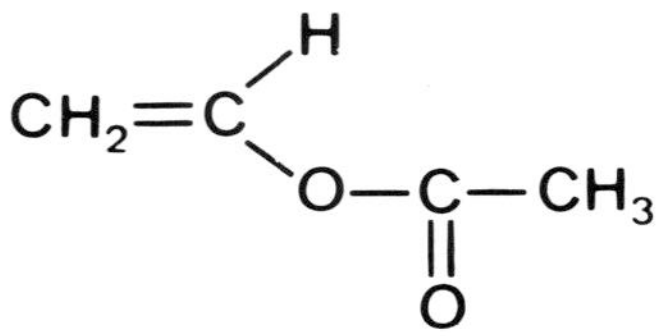

Acrylonitrile

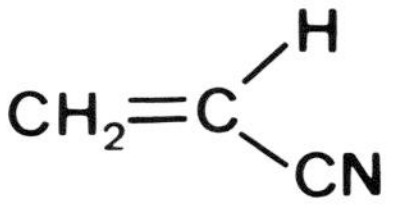

Styrene

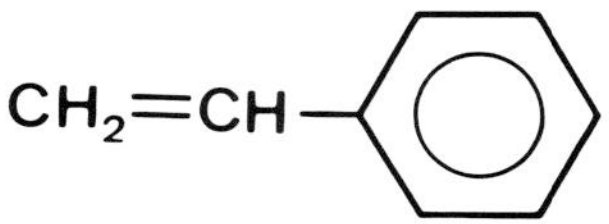

Cross-linkers

Ethylene glycol dimethacrylate derivatives
$CH_2=C(CH_3)COO(CH_2CH_2O)_nOCC(CH_3)=CH_2$

Methylene-bis-acrylamide
$CH_2=CHCONHCH_2NHOCCH=CH_2$

Polymeric cross-linker[44]

HEMA monomer. The monomer can be purified by first dissolving it in water, extracting with a nonpolar solvent such as carbon tetrachloride or hexane followed by alumina treatment to remove polar impurities, and finally distilling under vacuum in the presence of a polymerization inhibitor such as hydroquinone.

A. Bulk/Solution Polymerization

Hydrogels are commonly prepared by the bulk/solution polymerization process where a mixture of monomer, initiator, and an optional solvent, usually water, is polymerized in a mold to form sheets and cylinders.[8,14] The sample thickness for the bulk/solution polymerization process is generally kept small (millimeter range) because of the need to dissipate the heat of polymerization. Localized inhomogeneity and opaqueness are generally observed in thick, bulk-polymerized samples due to the lack of efficient dissipation of heat generated from the reaction exotherm.

Since oxygen inhibits the free-radical polymerization process, degassing of the monomer mixture and nitrogen purging are often needed for the polymerization process. Several studies have been reported in the literature by first incorporating drugs of interest into the monomer mixture followed by polymerization, thereby entrapping the drug within the hydrogel matrix. However, this is not a good way of making drug delivery systems because of the possible side reactions between the drug and the monomer and the inability to remove the potentially toxic residual monomer and initiator while keeping the entrapped drug intact. The preferable way is to polymerize the hydrogel in the absence of the drug. Subsequently, the hydrogel can be extracted in a good swelling solvent to remove the residual monomer, initiator, and other impurities. The drug loading can then be achieved by equilibrating the cleaned hydrogel in a concentrated drug solution prepared with a good swelling solvent. The swelling and loading characteristics of hydrogels will be addressed in more detail in the next section.

For drug delivery applications, the hydrogel sheets and cylinders prepared by the bulk polymerization process are useful only in topical and implant applications. To be useful for oral dosage forms, hydrogel would have to be in granules prepared by dicing or granulating hydrogel sheets or cylinders followed by sieving into proper particle sizes. Such hydrogel granules are generally irregular in shape which may be objectionable not only from the standpoint of product aesthetic, but also from the standpoint of reproducibility in controlling the drug release.

B. Suspension Polymerization

A more elegant and preferred way of preparing hydrogel particles is the suspension polymerization process which is capable of producing spherical hydrogel beads with particle size ranging from approximately 100 μm up to 5 mm in diameter.[15-21] The suspension polymerization technique has been used successfully in producing spherical beads of polystyrene, poly(vinylchloride), polyacrylates, and poly(vinyl acetate), mostly for the production of ion exchange resins.[15] Suspension polymerization is similar to bulk polymerization with respect to reaction kinetics; however, the monomer phase is dispersed by mechanical agitation into droplets in a nonsolvent medium with the aid of a protective colloid as a suspension stabilizer. The monomer droplets, which are generally larger than those of a true emulsion, are then polymerized through a heat-induced decomposition of the free radical initiator. The dissipation of heat evolved during polymerization is more efficient in suspension polymerization because the suspending medium, mostly water, acts as a heat sink. This process yields spherical beads with uniform composition in a one-step process. The hydrogel beads obtained from this process can easily be separated from the suspending aqueous phase and extracted with a good swelling solvent, usually ethanol or ethanol/water mixtures, before being used for drug loading. A typical batch of suspension-polymerized hydrogel beads is shown in Figure 1.

FIGURE 1. Photograph of a typical batch of suspension polymerized hydrogel beads.

The earliest references to the suspension polymerization of water-soluble monomers to form hydrogel beads are those by Speiser and co-workers in the late 1960s.[16-18] They prepared polymer beads by copolymerizing water-soluble acidic monomers such as α- and β-methacrylic acids with water-insoluble monomers such as methylmethacrylate and vinyl acetate. The suspension polymerization was carried out in a concentrated aqueous solution of sodium sulfate to decrease the solubility of water-soluble monomers in water. Water-soluble polymers such as carboxyvinyl polymers (Carbopol®) and polyvinylpyrrolidone were used as the suspension stabilizer. Azobisisobutyronitrile and benzoyl peroxide were used as the free radical initiator. Speiser and co-workers were able to prepare hydrogel beads of particle size ranging from 0.3 to 1.0 mm. However, most of their suspension polymerization was carried out in the presence of the drug (chloramphenicol and chlorothiazide), therefore rendering their systems not suitable for practical use because of the impossibility of cleaning the hydrogel beads without leaching out the drug.

Other references describe the suspension polymerization of 2-hydroxyethyl methacrylate using silicone oil, mineral oil, or xylene as the insoluble suspending phase.[19] The particles obtained from such processes are generally very irregular with rough surfaces.[21] The suspension polymerization of 2-hydroxyethyl methacrylate in the presence of 0.5 to 2% of short-chain cross-linking agent and using an aqueous salt solution as a suspending medium has been described by Kliment and co-workers;[20] however, there is no mention of the suspending agent.

More recently, an improved process for the preparation of spherical hydrogel beads in the presence of a polymeric cross-linking agent and a water-insoluble, gelatinous, strong water-bonding inorganic hydroxide as suspending agents in the absence of excess alkali has been reported by Mueller et al.[21] Therefore, in addition to the well-known water-soluble polymers such as polyvinylpyrrolidone, polyvinyl alcohol, and hydroxyethyl cellulose, insoluble inorganic salts such as calcium phosphate, calcium oxalate, calcium carbonate, calcium sulfate, and magnesium phosphate, as well as insoluble hydroxides such as magnesium hydroxide and aluminum hydroxide, can be used as efficient suspending agents for suspension polymerization. Mueller et al. reported that only the insoluble, gelatinous inorganic hydroxides such as magnesium hydroxide give rise to spherical beads with smooth surfaces when used in the suspension polymerization of their specific monomers. A typical example of the synthesis of their hydrogel beads is given below.

A 360-g, 20% (w/w) aqueous sodium chloride and 23 g of magnesium chloride hexahydrate were charged into a smooth-wall, 1-ℓ resin flask equipped with a baffle and anchor-type stirrer. A nitrogen blanket was maintained in the reaction flask at all times. The solution was heated slowly to 80°C with rapid stirring, and 123 mℓ of a 1N sodium hydroxide solution was then added dropwise to this solution to form a fine, gelatinous precipitate of magnesium hydroxide in the reaction flask.

After all of the sodium hydroxide was added, the stirring speed was reduced to 150 rpm and the monomer mixture (140 g of HEMA and 60 g of a polymeric cross-linking agent) containing a small amount of dissolved initiator (0.2 g of *t*-butyl peroctoate) was added. The reaction mixture was stirred under nitrogen at 150 rpm at 80°C for 3 hr. The temperature was then raised to 100°C for 1 hr, after which time the flask was cooled to room temperature, and 10 mℓ of concentrated HCl was then added to dissolve the magnesium hydroxide suspending agent. The reaction mixture was then filtered, washed with 2000 mℓ of water, and soaked overnight in 500 mℓ of ethanol to extract any residual monomer. Afterwards, the beads were isolated by filtration and dried.

IV. EQUILIBRIUM SWELLING PROPERTIES

Hydrogels can be considered as having two components, namely, a solid component of polymer network and a variable aqueous component which can undergo exchange with the environment. The network is held together mostly by covalent cross-links and, to a lesser extent, by physical cross-links such as chain entanglement and crystallites. The driving force of swelling is the water chemical potential difference between the network and external aqueous phase. In other words, the osmotic pressure due to the polymer segment in the gel network causes the swelling. Thermodynamically, the swelling properties of a hydrogel network are controlled by the combination of free energies of mixing between water (or any other solvent) and polymer chains and by the elastic response of the network to volume increase due to water absorption. At swelling equilibrium, the elastic response of the network exactly balances the difference in chemical potential of water between the swollen network and external aqueous phase. The magnitude of the equilibrium swelling can be related to the degree of cross-linking in the hydrogel network. One approach in the evaluation of cross-linking density is the use of tensile measurements.[22] However, this technique fails to take into account the effect of physical cross-links, such as chain entanglements, which tend to disentangle during extension. The other approach, which is preferable, is the equilibrium swelling experiments,[23-25] since the disentanglement process does not occur with swelling. Another useful parameter for the characterization of the cross-linked network structure in polymers is the molecular weight between cross-links, M_c.[11,23] Most of the previous swelling analyses have been based on the assumption of a loosely cross-linked network of macro-molecular chains exhibiting Gaussian distribution. Recently Peppas et al.[24,25] presented an accurate method for the determination of M_c for highly cross-linked PHEMA hydrogels.

The equilibrium water content (EWC) is generally affected by the nature of the hydrophilic monomers used in the hydrogel preparation, the nature and density of cross-links, and factors such as temperature, osmolarity, and pH of the hydrating medium. However, PHEMA has been reported to be little affected by variations in these factors; its EWC (about 40%) is only slightly dependent on the degree of cross-linking. This has been attributed to an unfavorable interaction parameter χ for the system PHEMA-water. The EWC of a hydrogel is normally defined as the weight of water in the polymer at equilibrium multiplied by 100% and divided by the total weight of the swollen polymer. This is perhaps one of the most important properties for hydrogels since it has a profound impact on the mechanical properties, permeability, surface properties, and biocompatibility of the hydrogel material. Co-polymerizing HEMA with hydrophobic monomers such as methacrylate or styrene reduces the EWC below that of PHEMA (40%), while copolymerizing HEMA with more hydrophilic monomers such as vinylpyrrolidone or methacrylic acid increases the EWC above that of PHEMA. Thus, hydrogels with EWC ranging from less than 10 to above 90% can be obtained by copolymerizing suitable combinations of hydrophobic and hydrophilic monomers.

Microscopically, dissolved drugs (or other solutes) can enter a swollen hydrogel if they are smaller than the interstitial space of mesh size of the network. The physical size and shape of the interstitial space in the network may change with time because of the constant thermal motion of the polymer segments; however, a statistically averaged mesh size may be calculated. Recent results indicate that, for cross-linked PHEMA, this mesh size is hardly more than 100 Å.[24] In general, one can expect in a qualitative sense that the average interstitial space increases as water content in the hydrogel increases.

As to the effect of equilibrium water content on the mechanism of drug permeation through hydrogel membranes, Yasuda et al.[26,27] derived a theoretical expression relating the solute diffusion coefficient in a water-swollen polymeric membrane to the free volume and the degree of hydration in the membrane. Conformity of experimental results to the theory suggests that the permeation of solute occurs predominately through the porous regions of the network. As pointed out by Yasuda et al., these porous, water-filled regions through which the transport of permeant can occur may only be conceived as fluctuating pores or channels of the polymer matrix which are not fixed either in size or in location. More recently, Zentner et al.[28,29] studied the effect of cross-linking agent on the progesterone permeation through swollen hydrogel membranes. In addition to the decrease in progesterone diffusion coefficient with increasing cross-linker level, they found that at low concentrations of cross-linker, the chain length of the cross-linker did not affect the "fluctuating pore" permeation mechanism. However, at high concentrations of cross-linker, the diffusion coefficient of progesterone in the system with a shorter cross-linker, EGDMA, was relatively independent of the cross-linker concentration. This was rationalized as a change in permeation mechanism to that of a solution-diffusion-controlled process. Such transition from porous to solution-diffusion transport is consistent with the water permeation results previously reported by Wisniewski et al.[30]

V. DRUG LOADING PROPERTIES

In order to develop a useful hydrogel drug delivery system, one has to achieve a sufficiently high drug loading in the polymer while at the same time providing a release rate consistent with biological requirements in terms of the magnitude and duration of drug release. From the safety standpoint, it is necessary to extract all leachable by-products and residual monomers left after polymerization in order to make hydrogels acceptable for drug delivery applications. Therefore, for all practical purposes, one is limited to polymerize hydrogels in the absence of the drug and to carry out drug loading with the extracted hydrogels by equilibrium absorption from concentrated drug solution. In this case, a high degree of swelling is needed to achieve a high drug loading, while a low to moderate degree of swelling in water is desired to control the rate of diffusional drug release.

The maximum swelling in a hydrogel network is found by measuring the equilibrium solvent content (defined similar to EWC) of the polymer for a series of solvents. If a hydrogel sample is allowed to swell in a series of solvents having a range of solubility parameters (defined as the square root of the cohesive energy densities of the solvents), the solubility parameter of the hydrogel network is found at the point of maximum swelling. A typical swelling curve for PHEMA is shown in Figure 2, where a series of ethanol/water mixtures were used as solvents.[31] The maximum degree of swelling (68%) for PHEMA apparently occurs at a solvent composition of 60% ethanol, giving rise to a solubility parameter for PHEMA at about 19.1 (in $[cal/cm^3]^{1/2}$) between that of ethanol (12.7) and water (23.4).

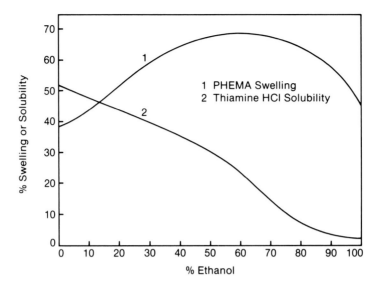

FIGURE 2.   Equilibrium solvent content of PHEMA and thiamine HCl solubility in ethanol/water mixtures.

When there are no substantial variations of drug solubilities over the solvent composition range, one can choose the solvent composition at maximum swelling as the loading solvent. A concentrated drug solution prepared in such solvent composition should allow a maximum amount of drug to be imbibed into the hydrogel at swelling equilibrium. Subsequent separation, rinsing, and drying will result in glassy, dehydrated hydrogel matrices containing uniformly dissolved or dispersed drug. The volume of the loading solution is generally greatly in excess of the volume of the hydrogel (e.g., larger than 10:1 ratio) in order to ensure that the external drug concentration remains relatively constant, even if a large partition effect exists. On the other hand, a substantial difference in drug solubility can exist over the swelling solvent compositions such as shown in Figure 2, where a nearly 20-fold difference is observed between thiamine HCl solubilities in water and ethanol. Under this situation, the solvent composition at maximum swelling may not be the optimum loading solvent for PHEMA because of the low drug solubility. Obviously, a compromise between swelling solvent composition and the corresponding drug solubility has to be made in order to maximize the loading capability. In fact, for the PHEMA-thiamine HCl system shown in Figure 2, 30% ethanol in water turned out to be the optimum loading solvent.

As to the solubility and partitioning effects which determine to a large extent the final drug loading, it is generally observed that the drug would be more soluble in the hydrogel phase as the difference in the solubility parameters of the drug and hydrogel becomes smaller. The partition coefficient, defined as the ratio of the drug concentration in the polymer to the equilibrium drug concentration in the external solvent medium, may also be concentration dependent. As an example, the absorption isotherms of thiamine HCl in PHEMA swollen in water as well as in 30:70 ethanol to water are illustrated in Figure 3 as a function of the drug concentration in solution,[32] in which the slope of the isotherm at any loading solution concentration determines the partition coefficient at that point. In this case, an approach to saturation accompanied by a diminishing partition coefficient is observed at higher drug concentrations. The construction of absorption isotherms similar to those shown in Figure 3 is a necessary step in selecting any hydrogel-drug combination for drug delivery applications. With the use of such absorption isotherms, one can easily determine the loading solution concentration and the appropriate loading solvent needed to achieve a specific drug loading.

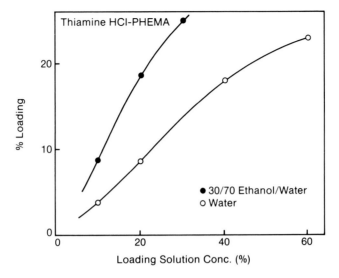

FIGURE 3. Absorption isotherms of thiamine HCl in water and ethanol/water-swollen PHEMA at 37.5°C. (From Good, W. R. and Lee, P. I., in *Medical Applications of Sustained Release*, Vol. 1, Langer, R. S. and Wise, D. L., Eds., CRC Press, Boca Raton, Fla., 1984, 1. With permission.)

VI. CHARACTERIZATION OF DIFFUSION PROPERTIES IN SWOLLEN HYDROGELS

Generally, the diffusion properties of solutes in polymers can be evaluated by either the membrane permeation method or the sorption/desorption method. In the former case, the permeation time-lag experiment has been widely used for the measurement of diffusion coefficient. However, this requires the elimination of boundary layer resistances and materials strong enough to be made into membranes. Hydrogels, being more fragile, especially when the equilibrium water content is high, may not be able to meet the latter requirement. Furthermore, for drugs with large molecular sizes and small diffusion coefficients, it often takes an impractically long time to establish the permeation time lag and steady state in a permeation experiment. Therefore, it is more convenient to characterize diffusion properties in swollen hydrogels using the sorption/desorption method.

The simplest way to obtain drug diffusion coefficients is to measure the initial rate of desorption of drug from a swollen hydrogel membrane or sheet of thickness l into an infinite volume. By plotting the fractional amount of drug desorbed, M_t/M_∞, vs. t, the drug diffusion coefficient, D, can be evaluated from the slope of the initial linear plot using the following equation:[33]

$$M_t/M_\infty = (4/l)\,[Dt/\pi]^{1/2} \qquad (1)$$

which is accurate to within 1% for up to approximately 60% of the total amount released.

When the hydrogel membrane or sheet contains dispersed drug, the desorption kinetics can be analyzed by the familiar Higuchi equation:[34]

$$M_t = [C_s(2A - C_s)Dt]^{1/2} \qquad (2)$$

which has been applied to drug delivery systems whenever the initial drug loading per unit volume, A, is greater than the drug solubility in the matrix, C_s. However, because of the pseudosteady-state assumptions involved, Higuchi's equation is only valid when the drug loading is in excess of the drug solubility ($A \gg C_s$). At the limit of $A \to C_s$, Higuchi's equation gives a result 11.3% smaller than the exact solution. Based on a refined integral method for moving boundary problems, Lee[35] recently presented a simple analytical solution for this problem which is uniformly valid over all A/C_s values:

$$M_t = C_s(1 + H)[Dt/3H]^{1/2} \qquad (3)$$

where

$$H = C_s^{-1}[5A + (A^2 - C_s^2)^{1/2}] - 4$$

When Equation 3 is applied to the present desorption system, the deviations from the exact results are consistently one order of magnitude smaller than those of Higuchi's equation. As $A/C_s > 1.04$, Equation 3 has an accuracy within 1% of the exact solution. Therefore, diffusion coefficients evaluated from desorption experiments using Equation 3 would be much more accurate than Equation 2, particularly at low A/C_s values. The latter situation occurs quite often in delivery systems involving hydrogels and drugs of high water solubility.

Another efficient method for the evaluation of transport coefficients pertinent to the design of hydrogel drug delivery systems is to follow the concentration change in a constant, finite volume as the drug is absorbed or desorbed by the swollen hydrogel material. This method has the distinct advantage that both the drug diffusion coefficient and partition coefficient can be obtained accurately from just one experiment with a simple apparatus and a wide range of accessible experimental conditions. Furthermore, the sample geometry and mechanical strength are not restricted in such an experiment. The exact solution for this diffusion system has been known for some time; however, all these solutions require either a laborious construction of master plots of calculated values or a numerical data-fitting routine when they are applied to the analysis of experimental data.

A set of simple, accurate approximate solutions for different geometries recently given by Lee[36] is much more convenient to use than the existing solutions. It enables one to evaluate the diffusion coefficient from simple algebraic equations based on sorption data. For an absorption experiment, the equation for flat sheet configuration is

$$F(C) = [(C_o^2 - C^2)/2C^2 + ln(C/C_o)](3\lambda^2/16) = Dt/l^2 \qquad (4)$$

where C_o is the original drug concentration in the finite volume of external solution, C the corresponding concentration at time t during the absorption, and l the thickness of the hydrogel sheet. The effective volume ratio is defined as:

$$\lambda = C_\infty/(C_o - C_\infty) = V/SKl \qquad (5)$$

with V being the total liquid volume, S the area of each side of the sheet, K the partition coefficient, and C the equilibrium drug concentration in the external solution after the completion of the absorption experiment. The concentration change is further related to the fraction absorbed by:

$$C/C_o = 1 - [1/(1 + \lambda)](M_t/M_\infty) \qquad (6)$$

Therefore, in a finite volume absorption experiment, the monotonic decrease of drug concentration in the solution external to a hydrogel material fabricated in sheet form is followed as a function of time. The left-hand side of Equation 4 is then calculated and

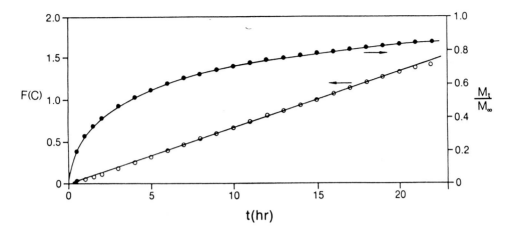

FIGURE 4.   Application of Equation 4 to the absorption of dexamethasone in a PHEMA disk from a constant finite volume at pH 7 and 37°C; k = 51.1, λ = 0.465, giving D = 1.18 × 10⁻⁸ cm²/sec. (From Lee, P. I., in *Controlled Release of Bioactive Materials,* Baker, R., Ed., Academic Press, New York, 1980, 135. With permission.)

plotted against time. From the slope of such a linear plot, the diffusion coefficient can be evaluated. The partition coefficient of the drug in the hydrogel can also be calculated from the same set of data using Equation 5. A typical application of Equation 4 is shown in Figure 4. A similar set of equations for spheres and cylinders may also be found.[36] Another set of equations is also applicable to desorption experiments; however, in this case, C has to be replaced by $(C_s - KC)$, C_o by C_s, and C by $(C_s - KC_\infty)$ in Equations 4 through 6, where C_s is the drug solubility in the polymer and all other parameters have been previously defined. The only limitation in the applications of these equations for the analysis of sorption data is that λ has to be smaller than 10. For λ values greater than 10, which correspond to experimental conditions of near-constant external concentrations, the solution for constant external concentration such as Equation 1 should be used for the evaluation of drug diffusion coefficient.

VII. KINETICS OF SWELLING AND DRUG RELEASE FROM DRY HYDROGELS

In many applications, especially oral delivery, drug-loaded hydrogels are usually stored in a dry, glassy state before usage due to stability and dosing requirement. The release of water-soluble drugs from initially dry hydrogel matrices generally involves the simultaneous absorption of water and desorption of drug via a swelling-controlled mechanism.[5] Thus, as water penetrates a glassy hydrogel matrix containing dissolved or dispersed drug, the polymer swells and its glass transition temperature is lowered. In most cases, a sharp penetrating solvent front separating the glassy from the rubbery phase, in addition to a volume swelling, is also observed. A typical example is shown in Figure 5. In terms of drug distribution, this solvent front also separates the undissolved core from the partially extracted region, with the dissolved drug diffusing through this swollen rubbery region into the external releasing medium. Depending on the relative magnitude of the rate of polymer relaxation at the penetrating solvent front and the rate of diffusion of the dissolved drug, the release behavior during the initial stage of the solvent penetration may range from Fickian to non-Fickian (anomalous), including the so-called Case II diffusion. Typically, for a polymer slab, Fickian diffusion is characterized by a square-root-of-time dependence in both the amount diffused and the penetrating diffusion front position. On the other hand, Case II transport, which is

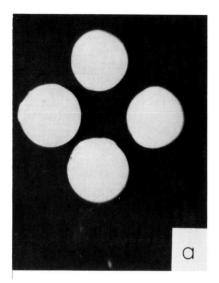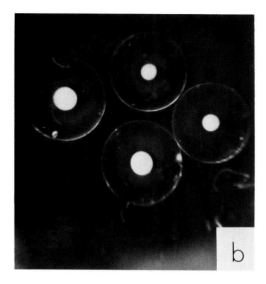

FIGURE 5.   Photographs showing typical solvent penetration behavior in drug-loaded glass hydrogel beads: (a) original, and (b) solvent penetrated. (From Lee, P. I., *J. Pharm. Sci.*, 73, 1344, 1984. Reproduced with permission of the copyright owner, the American Pharmaceutical Association.)

completely governed by the rate of polymer relaxation, exhibits a linear-time dependence in both the amount diffused and the penetrating front position. In most cases, the intermediate situation, which is often termed non-Fickian or anomalous diffusion, will exist whenever the rates of Fickian diffusion and polymer relaxation are comparable.

When the fractional drug release from an initially dry hydrogel sheet is plotted as a function of the square root of time as shown in Figure 6 for thiamine HCl release from PHEMA, linearity in the plot is observed only after longer periods of time. This illustrates the non-Fickian and time-dependent nature of the initial swelling period. Once the hydrogel matrix is hydrated, the drug release becomes Fickian, giving rise to the linearity in Figure 6 at long periods of time. Phenomenologically, it is possible to express the fraction released, M_t/M_∞, as a power function of time t, for at least the short time period,

$$M_t/M_\infty = kt^n \qquad (7)$$

where k is a constant characteristic of the system and n is an exponent characteristic of the mode of transport. For n = 0.5, the solvent diffusion or drug release follows the well-known Fickian diffusion mechanism. For n > 0.5, non-Fickian or anomalous diffusion behavior is generally observed. The special case of n = 1 gives rise to a Case II transport mechanism, which is of particular interest because the drug release from such devices having constant geometry will be zero order. Other parameters such as the so-called Deborah number,[37] which measures the relative importance of relaxation to diffusion, and the swelling interface number,[38] which compares the relative mobilities of the penetrating solvent and the drug in the presence of polymer relaxation, are valuable in the conceptual realization of various diffusion mechanisms. However, very limited experimental determinations of these parameters have been reported.

Only a few attempts have been made to model the swelling-controlled release systems. Good[7] employed a time-dependent drug diffusion coefficient which was set to be proportional to the fractional solvent absorption. The results were used to fit experimental diffusion data for drug release from initially dry hydrogels. Lee[35,39] analyzed drug release from polymer

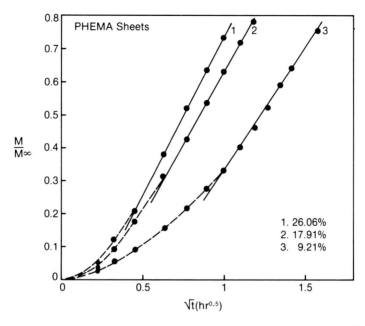

FIGURE 6. Effect of loading level on the fractional release of thiamine HCl from initially dry PHEMA sheets at 37.5°C plotted as a function of square root of time.

matrices involving moving boundaries generated by both the polymer swelling and erosion. Accurate approximate analytical solutions for various geometries were presented. Peppas et al.[40] developed a pseudosteady-state model for drug diffusion from polymers with volume expansion. Korsmeyer and Peppas[41] developed mathematical models based on a drug diffusion coefficient which depends on the concentration of absorbed solvent in a functional form consistent with the free-volume theory. Recently, Lee[5] demonstrated that by incorporating a time dependence explicitly into the drug diffusion coefficient to reflect the polymer relaxation process, various release behaviors from hydrogel matrices ranging from Case II to Fickian can be described by the analytical solutions to the corresponding moving boundary problem formulated for a swellable dispersed system. The predicted release behavior is consistent with physical observations and the Deborah number concept.

VIII. DIMENSIONAL CHANGES DURING DRUG RELEASE

The relaxation and diffusion processes associated with the dynamic swelling of a glassy hydrogel are often reflected in the change of sample dimension as a function of time. The sorption kinetics and the transient dimensional changes during the release of a dissolved or dispersed drug from an initially dehydrated hydrogel are more complex than the case of single penetrant transport in glassy polymers. Generally, they are significantly affected by the local drug concentration. The presence of a water-soluble drug alters both the swelling osmotic pressure and the associated time-dependent relaxation of the hydrogel network during the simultaneous absorption of water and desorption of drug. A detailed investigation on the dimensional changes during the entire course of the simultaneous water penetration and release of thiamine HCl from glassy PHEMA beads was first reported by Lee.[42]

A typical example of the effect of loading level on the fractional release of thiamine HCl from glassy PHEMA beads at 37.5°C is shown in Figure 7. The associated solvent front penetration behavior is presented as a plot of the square root of time in Figure 8. The

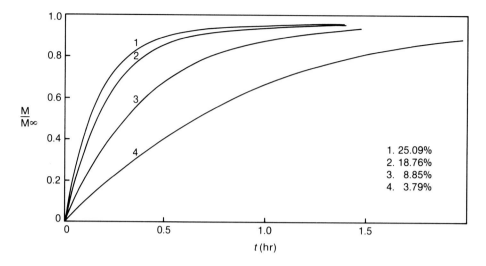

FIGURE 7.   Effect of loading level on the fractional release of thiamine HCl from initially dry PHEMA beads at 37.5°C. (From Lee, P. I., *Polym. Commun.*, 24, 45, 1983. With permission.)

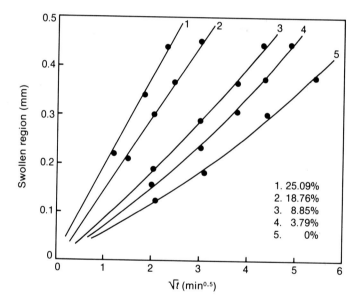

FIGURE 8.   Effect of loading level on the solvent front penetration as a function of square root of time during the release of thiamine HCl from initially dry PHEMA beads at 37.5°C. (From Lee, P. I., *Polym. Commun.*, 24, 45, 1983. With permission.)

fractional release and solvent front penetration are observed to behave more Fickian as the thiamine HCl loading level increases. Such a transition can be considered as a change of relative importance of the diffusion process vs. the polymer relaxation as a function of drug loading.

The transient dimensional changes during the simultaneous solvent penetration and thiamine HCl release from PHEMA beads having various drug loading levels are shown in Figure 9. The spherical geometry utilized has the advantage of eliminating the anisotropy and the edge effects normally associated with dimensional measurements in glass polymers.

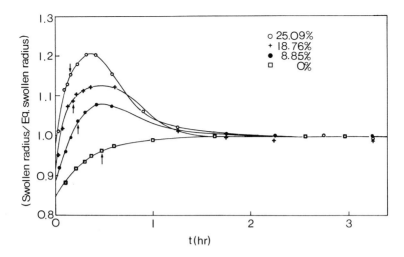

FIGURE 9.   Effect of loading level on the relative dimensional changes with respect to the equilibrium swollen dimension during the release of thiamine HCl from initially dry PHEMA beads at 37.5°C.[42]

As shown in Figure 9, the radius of an unloaded bead increases monotonically toward the equilibrium radius, whereas that of a loaded bead goes through a maximum corresponding to approximately 70 to 80% total release before reaching the final equilibrium value. The continuous increase of the radius after the penetrating fronts have met is believed to be primarily the result of the solvent concentration gradient behind the swelling fronts. The reference state in Figure 9 is the equilibrium swollen radius; therefore the plots converge to the same equilibrium value, 1.

The presence of homogeneously dissolved or dispersed thiamine HCl in PHEMA beads provides an additional osmotic driving force which alters both the total swelling osmotic pressure and the associated time-dependent relaxation of the hydrogel network during the simultaneous sorption of water and desorption of drug. The sorption of water tends to increase, whereas the desorption of drug tends to decrease, the dimension of the hydrogel bead. The combination of these two competing processes results in the observed maximum in the transient dimensional changes for the drug-loaded hydrogel beads.

IX. METHODS OF MODIFYING RELEASE KINETICS

Diffusion-controlled matrix systems where the drug is uniformly dissolved or dispersed in a polymer matrix generally exhibit release rates continuously diminishing with time. This is the result of the increasing diffusional resistance and decreasing area at the penetrating diffusion front. Hydrogel matrices show no exception from this, despite the theoretical prospect of having a totally relaxation-controlled (Case II) situation, thereby achieving zero-order release. Experimentally, a hydrogel having pure Case II swelling kinetics is yet to be demonstrated. Even if this is done, the deviation from Case II behavior at higher drug loading levels as shown in the previous section still complicates the matter.

Methods to modify the kinetics of drug release from a monolithic hydrogel matrix were mainly in the area of rate-controlling membranes. When a hydrogel matrix is used as a drug reservoir, a rate-controlling membrane can be formed by a newly developed interpenetrating network (IPN) technique, where the surface layer of the matrix is first imbibed with a monomer solution followed by either a free radical polymerization induced by heat or UV radiation as shown by Lee et al.[43] or an *in situ* polycondensation generated by immersion

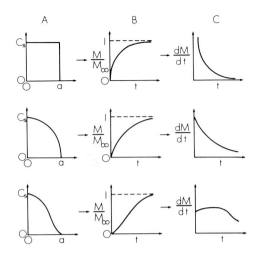

FIGURE 10. Theoretical profiles illustrating the effect of initial drug concentration distribution on the characteristics of drug release from spherical matrices: (A) concentration profile, (B) cumulative release, and (C) release rate. (From Lee, P. I., *Polymer*, 25, 973, 1984, by permission of the publishers, Butterworth & Co. (Publishers), Ltd.©)

in a second reactant solution to arrive at a less permeable, rate-controlling membrane layer as reported by Mueller and Heiber.[44]

Recently, Lee[4,45] demonstrated a novel approach to constant-rate drug release from glassy hydrogel beads via an immobilized nonuniform drug concentration distribution. The results indicate that a constant rate of drug release can be achieved via a sigmoidal type of drug concentration distribution without the need to have a saturated drug reservoir. The effect of nonuniform initial drug distribution on the release behavior has not been discussed in the literature prior to Lee's work. Figure 10 illustrates the characteristics of drug release from spherical matrices as a function of the initial drug distribution. Based on theoretical solutions, Figure 10 shows that both the uniform and the parabolic initial concentration distribution result in an initially high rate of release followed by a rapid decline; the latter has a reduced rate of release compared with the former. In contrast, a sigmoidal initial distribution is capable of introducing a characteristic inflection point and, therefore, considerable linearity into the cumulative release curve. As a result, a prolonged constant rate of drug release similar to a membrane reservoir system is obtained. The parabolic type of concentration distribution is characteristic of Fickian diffusion in rubbery polymers, whereas the sigmoidal distribution is characteristic of glassy polymers partially penetrated by a swelling solvent undergoing non-Fickian diffusion.

Based on the above considerations, Lee developed a controlled-extraction process to partially penetrate and extract drug-loaded hydrogel beads with a swelling solvent. The non-Fickian diffusion behavior enables the development of a sigmoidal concentration profile for both the drug and the solvent. Immediately after separating the extracting solvent, the controlled release beads were freeze-dried under vacuum to rapidly remove the swelling solvent and to immobilize a nonuniform, sigmoidal drug distribution. This process was applied to cross-linked HEMA copolymer hydrogel beads loaded with a very water-soluble drug, oxprenolol HCl. It is usually very difficult to achieve a zero-order release from a membrane-reservoir system because of this drug's high water solubility.

The existence of immobilized sigmoidal drug concentration distribution was confirmed

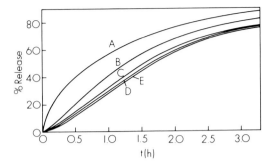

FIGURE 11. Effect of controlled extraction time in water on the in vitro release of oxprenolol HCl from hydrogen beads: (A) 0 min, (B) 5 min, (C) 15 min, (D) 20 min, and (E) 30 min. (From Lee, P. I., *Polymer,* 25, 973, 1984, by permission of the publishers, Butterworth & Co. (Publishers), Ltd.©)

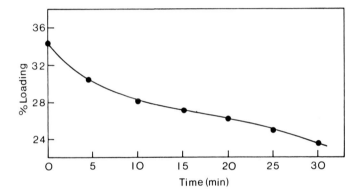

FIGURE 12. Oxprenolol HCl loading as a function of controlled extraction time in water. (From Lee, P. I., *J. Pharm. Sci.,* 73, 1344, 1984. Reproduced with permission of the copyright owner, the American Pharmaceutical Association.)

using a scanning electron microscope X-ray chlorine scan technique. The corresponding in vitro oxprenolol HCl release behavior from controlled-extracted beads as compared with that of unextracted control is given in Figure 11. It is clear that a release time lag and a constant-release region similar to that of membrane-reservoir devices have been introduced by the process. With the increase in controlled release extraction time, the constant-rate release region can be extended. The constant release region also shows a progressively decreasing rate with increasing extraction time. Inevitably, a certain amount of drug will be lost during the controlled-extraction process. However, as shown in Figure 12, only up to 10% of the drug loading is removed. Stability tests on the system indicate that in the absence of moisture, the constant-release characteristics can be preserved for long periods of time as evidenced in Figure 13. The diffusion of entrapped drug does not occur until the hydrogel matrix is swollen at the time of usage.

In addition to being able to achieve a constant rate of drug release without the needing a saturated reservoir and a rate-controlling membrane, this process of controlled extraction provides a rational and practical way of modifying the release kinetics from hydrogel mat-

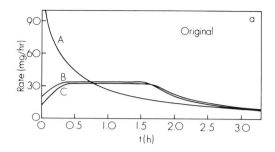

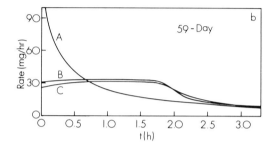

FIGURE 13.   Effect of storage time on the in vitro oxprenolol HCl release from hydrogel beads: (A) loaded control, (B) controlled extracted in water for 20 min, and (C)
controlled extracted in water for 30 min. (From Lee, P. I.,
Polymer, 25, 973, 1984, by permission of the publishers,
Butterworth & Co. (Publishers), Ltd.©)

rices. Hydrogels are particularly suitable for this application because they are capable of
immobilizing any nonuniform drug distribution introduced prior to the dehydration step, due
to its glassiness in the dry state.

REFERENCES

1. **Wichterle, O. and Lim, D.,** Hydrophilic gels for biological use, *Nature (London),* 185, 117, 1960.
2. **Ratner, B. D. and Hoffman, A. S.,** Synthetic hydrogels for biomedical applications, in *Hydrogels for Medical and Related Applications,* ACS Symp. Ser. 31, Andrade, J. D., Ed., American Chemical Society, Washington, D.C., 1976, 1.
3. **Pedley, D. G., Skelly, P. J., and Tighe, B. J.,** Hydrogels in biomedical applications, *Br. Polym. J.,* 12, 99, 1980.
4. **Lee, P. I.,** Novel approach to zero-order drug delivery via immobilized nonuniform drug distribution in glassy hydrogels,, *J. Pharm. Sci.,* 73, 1344, 1984.
5. **Lee, P. I.,** Kinetics of drug release from hydrogel matrices, *J. Controlled Release,* 2, 1985.
6. **Lee, P. I.,** Zero order release from glassy hydrogel beads, in Proc. 10th Int. Symp. Controlled Release of Bioactive Materials, Controlled Release Society, Lincolnshire, Ill., 1983, 136.
7. **Good, W. R.,** Diffusion of water soluble drugs from initially dry hydrogels, in *Polymeric Delivery Systems,* Kostelnik, R., Ed., Gordon and Breach, New York, 1976, 139.
8. **Wichterle, O.,** Hydrogels, in *Encyclopedia of Polymer Science and Technology,* Vol. 15, Mark, H. F. and Gaylord, N. G., Eds., Interscience, New York, 1971, 273.
9. **Yocum, R. H. and Nyquist, E. B.,** *Functional Monomers* Vols. 1 and 2, Marcel Dekker, New York, 1973.
10. **Mueller, K. F. and Good, W. R.,** U.S. Patent 4,177,056, 1979.

11. **Good, W. R. and Mueller, K. F.,** A new family of monolithic hydrogels for controlled release applications, in *Controlled Release of Bioactive Materials,* Baker, R., Ed., Academic Press, New York, 1980, 155.
12. **Graham, N. B. and McNeill, M. E.,** Hydrogels for controlled drug delivery, *Biomaterials,* 5, 27, 1984.
13. **Gregonis, D. E., Chean, C. M., and Andrade, J. D.,** The chemistry of some selected methacrylate hydrogels, in *Hydrogels for Medical and Related Applications,* ACS Symp. Ser. 31, Andrade, J. D., Ed., American Chemical Society, Washington, D.C., 1976, 88.
14. **Refojo, M. F.,** Hydrophobic interaction in poly(2-hydroxyethyl methacrylate) homogeneous hydrogel, *J. Polym. Sci. Polym. Chem. Ed.,* 5, 3103, 1967.
15. **Farber, E.,** Suspension polymerization, in *Encyclopedia of Polymer Science and Technology,* Vol. 13, Mark, H. F. and Gaylord, N. G., Eds., Interscience, New York, 1970, 552.
16. **Khanna, S. C., Jecklin, T., and Speiser, P.,** Bead polymerization technique for sustained-release doasage form, *J. Pharm. Sci.,* 59, 614, 1970.
17. **Speiser, P.,** U.S. Patent 3,390,050, 1968.
18. **Khanna, S. C. and Speiser, P.,** In-vitro release of chloroamphenicol from polymer beads of α-methacrylic acid and methyl methacrylate, *J. Pharm. Sci.,* 59, 1398, 1970.
19. U.S. Patents 3,567,118; 3,574,628; 3,575,123; 3,577,518; 3,583,957.
20. **Kliment, K., Vacik, J., Ott, Z., Majkus, V., Stol, M., Stoy, V., and Wichterle, O.,** U.S. Patent 3,689,634, 1972.
21. **Mueller, K. F., Heiber, S. J., and Plankl, W. L.,** U.S. Patent 4,224,427, 1980.
22. **Janacek, J. and Hasa, J.,** Structure and properties of hydrophilic polymers and their gels. VI. Equilibrium deformation behavior of PHEMA and PHEEMA networks prepared in the presence of a diluent and swollen with water, *Collect. Czech. Chem. Commun.,* 31, 2186, 1966.
23. **Migliaresi, C., Nicodemo, L., Nicolais, L., and Passerin, P.,** Physical characterization of PHEMA gels, *J. Biomed. Mater. Res.,* 15, 307, 1981.
24. **Peppas, N. A., Moynihan, H. J., and Lucht, L. M.,** The structure of highly crosslinked poly(2-hydroxyethyl methacrylate) hydrogels, *J. Biomed. Mater. Res.,* 19, 397, 1985.
25. **Barr-Howell, B. D. and Peppas, N. A.,** Importance of junction functionality in highly crosslinked polymers, *Polym. Bull.,* 13, 91, 1985.
26. **Yasuda, H., Lamaze, C. E., and Peterlin, A.,** Diffusive and hydraulic permeabilities of water in water-swollen polymer membranes, *J. Polym. Sci. Part A-2,* 9, 1117, 1971.
27. **Yasuda, H., Lamaze, C. E., and Ikenberry, L. D.,** Permeability of solutes through hydrated polymer membranes. I. Diffusion of sodium chloride, *Makromol. Chem.,* 118, 19, 1968.
28. **Zentner, G. M., Cardinal, J. R., and Kim, S. W.,** Progestin permeation through polymer membranes. I. Diffusion studies in plasma-soaked membranes, *J. Pharm. Sci.,* 69, 1347, 1978.
29. **Zentner, G. M., Cardinal, J. R., and Kim, S. W.,** Progestin permeation through polymer membranes. II. Diffusion studies on hydrogel membranes, *J. Pharm. Sci.,* 69, 1352, 1978.
30. **Wisniewski, S. J., Gregonis, D. E., Kim, S. W., and Andrade, J. D.,** Diffusion through hydrogel membranes, in *Hydrogels for Medical and Related Applications,* ACS Symp. Ser. 31, Andrade, J. D., Ed., American Chemical Society, Washington, D.C., 1976, 80.
31. **Lee, P. I.,** unpublished data.
32. **Good, W. R. and Lee, P. I.,** Membrane reservoir controlled drug delivery systems, in *Medical Applications of Sustained Release,* Vol. 1, Langer, R. S. and Wise, D. L., Eds., CRC Press, Boca Raton, Fla., 1984, 1.
33. **Crank, J.,** *Mathematics of Diffusion,* 2nd ed., Clarendon Press, 1975.
34. **Higuchi, T.,** Rate of release of medicaments from ointment bases containing drugs in suspension, *J. Pharm. Sci.,* 50, 874, 1961.
35. **Lee, P. I.,** Diffusional release of a solute from a polymer matrix — approximate analytical solutions, *J. Membr. Sci.,* 7, 255, 1980.
36. **Lee, P. I.,** Determination of diffusion coefficients by sorption from a constant, finite volume, in *Controlled Release of Bioactive Materials,* Baker, R., Ed., Academic Press, New York, 1980, 135.
37. **Vrentas, J. S., Jarzebski, C. M., and Duda, J. L.,** A Deborah Number for diffusion in polymer-solvent systems, *AIChE J.,* 21, 894, 1975.
38. **Korsmeyer, R. W. and Peppas, N. A.,** Macromolecular and modeling aspects of swelling-controlled systems, in *Controlled Release Delivery Systems,* Roseman, T. J. and Mansdorf, S. Z., Eds., Marcel Dekker, New York, 1983, 77.
39. **Lee, P. I.,** Controlled drug release from polymeric matrices involving moving boundaries, in *Controlled Release of Pesticides and Pharmaceuticals,* Lewis, D. H., Ed., Plenum Press, New York, 1981, 39.
40. **Peppas, N. A., Gurny, R., Doelker, E., and Buri, P.,** Modeling of drug diffusion through swellable polymeric systems, *J. Membr. Sci.,* 7, 241, 1980.
41. **Korsmeyer, R. W. and Peppas, N. A.,** Modeling drug release from swellable systems, in Proc. 10th Int. Symp. Controlled Release of Bioactive Materials, Controlled Release Society, 1983, 141.

42. **Lee, P. I.,** Dimensional changes during drug release from a glassy hydrogel matrix, *Polym. Commun.,* 24, 45, 1983.
43. **Lee, E. S., Kim, S. W., Kim, S. H., Cardinal J. R., and Jacob, H.,** Drug release from hydrogel devices with rate-controlling barriers, *J. Membr. Sci.,* 7, 293, 1980.
44. **Mueller, K. F. and Heiber, S. J.,** Gradient-IPN-modified hydrogel beads: their synthesis by diffusion-polycondensation and function as controlled drug delivery agents, *J. Appl. Polym. Sci.,* 27, 4043, 1982.
45. **Lee, P. I.,** Effect of non-uniform initial drug concentration distribution on the kinetics of drug release from glassy hydrogel matrices, *Polymer.* 25, 973, 1984.

Chapter 5

BIOERODIBLE POLYMERS FOR CONTROLLED RELEASE SYSTEMS

Howard B. Rosen, Joachim Kohn, Kam Leong, and Robert Langer

TABLE OF CONTENTS

I. INTRODUCTION

Other chapters in this book have discussed the controlled release of drugs from inert polymers. The use of inert polymers has limitations, however, since a subcutaneously implanted device requires surgery for removal (after the drug supply is depleted from the matrix) as well as for insertion. As a result, the development of a bioerodible drug delivery matrix for the controlled release of drugs is of interest since such a device obviates the need for its surgical removal.

Bioerodible polymers are often described as biodegradable in the environmental control literature and bioabsorbable in the medical literature. In general, the three adjectives are used interchangeably to describe synthetic or natural polymers which hydrolyze in the living organism.

Both bioerodible reservoir and matrix devices can be used. In a reservoir system, a core of drug is surrounded by a polymer and diffusion of the drug through the polymer is the rate-limiting step. In an idealized device, drug release is zero order as long as the drug concentration in the reservoir remains constant.[1] The rate of drug delivery from reservoir devices that have a rate-controlling bioerodible membrane surrounding a drug core is constant and predictable provided the membrane erodes long after drug delivery has been completed. If the rate-controlling membrane erodes significantly during drug delivery, changes in membrane thickness or physical properties would be reflected in changes in drug delivery rate. Thus, in this type of device the rate of drug release can be controlled by changing the nature of the bioerodible membrane; bioerosion serves to remove the expended device.

A matrix system has drug dissolved or dispersed uniformly throughout a solid polymer. Drug release from matrix devices can be controlled by either diffusion or erosion. If erosion of the matrix is much slower than diffusion, the release kinetics can be described by standard diffusion equations.[1] In an idealized, inert, flat, spherical, or cylindrical matrix, drug release is not zero order, but rather is linear with respect to the square root of the time (except at long intervals).[1] If, however, the drug is immobilized in the matrix so that diffusional release is minimal compared to erosion, the rate of drug release will be erosion controlled.

On a molecular level, there are three general mechanisms for polymer hydrolysis[2] (see Figure 1). Mechanism I describes the erosion of cross-linked polymers with hydrolytically unstable cross-links. As the cross-links are hydrolyzed, polymer chains are freed from the bulk matrix. These systems may be useful for drugs with low water solubility or macro-molecules which can be physically entangled in a cross-linked matrix so that they cannot easily diffuse outward. Mechanism II applies to water-insoluble polymers which solubilize as the result of hydrolysis, ionization, or protonation of a side group. In this type of erosion, the only reaction is solubilization and there is no significant change in polymer molecular weight. Consequently, such polymers are not generally useful for systemic applications because of difficulty in eliminating such molecules. These polymers may be useful, however, in topical or oral applications. Mechanism III is proposed for the erosion of insoluble polymers with labile backbone bonds. Hydrolysis causes scission of the polymer backbone which produces low molecular weight, water-soluble molecules. Because these polymers are converted to small water-soluble molecules, these systems are most useful for systemic administration of therapeutic agents from subcutaneous, intramuscular, and intraperitoneal implantation sites. Situations in which combinations of these mechanisms operate can be envisioned. For example, a cross-linked polymer may initially degrade to insoluble polymer chains (mechanism I) which are hydrolyzed further by backbone cleavage (mechanism III).[2]

These hydrolysis mechanisms, when considered macroscopically, result in two general erosion mechanisms — heterogeneous and homogeneous erosion.[2] Heterogeneous erosion occurs when hydrolysis takes place only at the surface of a polymer sample and is characterized by a sample which maintains its physical integrity as it degrades. In contrast,

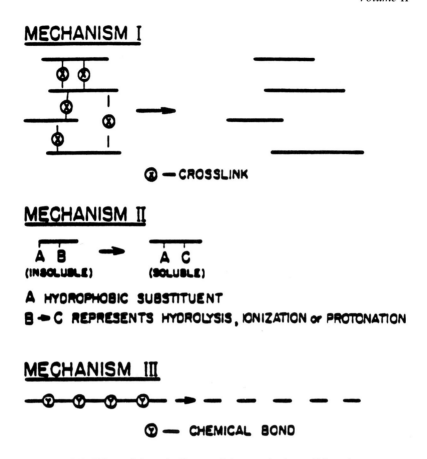

FIGURE 1.   Schematic diagram of three mechanisms of bioerosion.

homogeneous erosion is the result of hydrolysis occurring at an even rate throughout the polymer sample. Homogeneous erosion results in four stages of strength and integrity loss:[3]

1.   Hydration, resulting in the disruption of van der Waals forces and hydrogen bonds
2.   The initial breaking of covalent bonds, causing irreversible strength loss
3.   Mass loss, resulting from continuation of bond cleavage which is accompanied by the diffusion of low molecular weight hydrolysates away from the sample
4.   Complete absorption

In both heterogeneous and homogeneous erosion, mechanical strength deterioration may also be caused by absorption of compounds other than water such as lipids, steroids, and amino acids from the surrounding medium.[4]

Homogeneous and heterogeneous mechanisms are extreme cases and actual erosion commonly occurs by a combination of the two. The dominant erosion mechanism can be predicted from polymer hydrophobicity and morphology. Hydrophobic polymers are more likely to erode heterogeneously since water is excluded. Hydrophilic polymers absorb water so homogeneous erosion will be favored. Superimposed on the polymer hydrophobicity is polymer morphology. Most polymers are semicrystalline with crystalline domains separated by amorphous regions. The crystalline regions exclude water so more crystalline polymers tend towards heterogeneous erosion. Amorphous regions which are glassy absorb more water than crystalline regions, but less than amorphous regions which are rubbery.

Another important property of a bioerodible matrix is the biocompatibility of the polymer[5]

Table 1
SOME BIOERODIBLE POLYMERS

	Ref.
Polyamides	
Albumin[a]	12
Poly[(hydroxyalkyl)-L-glutamines][a]	89
Collagen[a]	11
Poly(L-leucine-co-L-aspartic acid)	13
Polydepsipeptide	90
Poly(β-alanine)	91
Poly(proline-co-glutamic acid)	92
Poly(gelatin-co-lysine esters)[a]	93
Poly(L-glutamic acid-co-γ-ethyl-L-glutamate)[a]	15
Polyesters	
Poly(β-propiolactone)	94
Poly(β-hydroxybutyrate)	95
Poly(lactic acid)[a]	22
Poly(glycolic acid)[a]	22
Poly(ϵ-caprolactone)[a]	55
Poly(alkelyene oxalates)[a]	96
Polydioxanone[a]	102
Poly(alkelyene diglycolates)[a]	97
Polyaminotriazole	89
Polydihydropyrans[a]	99
Poly(alkyl 2-cyanoacrylates)[a]	80
Poly(orthoesters)[a]	76
Polyanhydrides[a]	62
Chitosan	100
Polyurethanes	101
Polyacetals	68

[a]   Use in drug release systems reported.

and its erosion products.[6] The erosion products must either be safely eliminated from the body or metabolized to nontoxic substances.

Examples of the above general characteristics for specific bioerodible polymers are presented in the next section with particular attention to the synthesis procedures and formulation methods used. A final section discusses the modeling of drug release kinetics from bioerodible polymers.

II. BIOERODIBLE POLYMERS

The classification of a polymer as bioerodible is somewhat arbitrary. For example, even an "inert" polymer, polyethylene terephthalate, degrades in humans and dogs in 30 ± 7 years.[7] For controlled release applications, devices which range in life from 1 day to several years are desired. There are many synthetic and natural polymers which are potentially applicable to bioerodible drug delivery systems (see Table 1). This paper focuses on many of the major classes of erodible polymers.

A. Polyamides
Polyamides have received considerable attention and both synthetic and natural polyamides have been extensively investigated for their possible use as bioerodible polymers in biomedical applications. In spite of the high chemical inertness of the amide group (as compared, for example, to the ester group), polyamides are usually expected to be biodegradable,

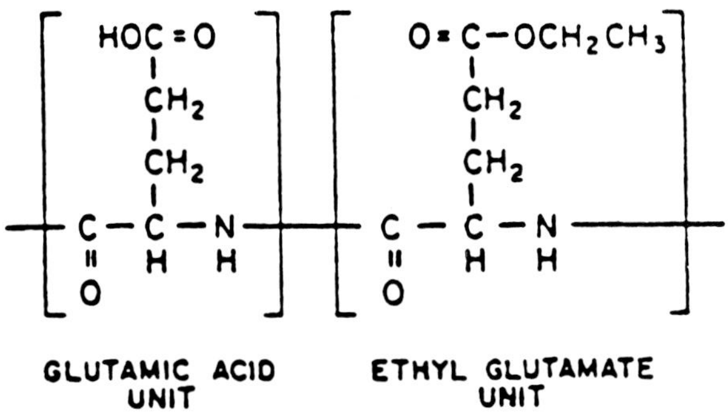

FIGURE 2. Poly(γ-ethyl-L-glutaminate-co-L-glutamic acid) structure.

because of the general assumption that the amide linkage would be susceptible to cleavage by nonspecific amidases. This view is supported by the reported failure of nylon prostheses,[8,9] which show surface cracks soon after implantation, followed by slow bulk erosion and loss of tensile strength.

Reconstituted collagen is probably one of the most widely used natural polymers for biomedical applications. It has been used as a suture, prosthesis, or wound dressing and it is degraded in vivo by the enzyme collagenase. Attempts to use reconstituted collagen as a basis for drug delivery systems have been only partly successful. When used for the delivery of pilocarpine[10] or medroxyprogesterone,[11] the release rates were not constant or reproducible. Similar problems were encountered when cross-linked serum albumin microbeads were used for the short-term delivery of progesterone.[12] It seems that unfavorable release kinetics are often observed when natural biopolymers are used for drug delivery applications. Therefore an increasing research effort is directed toward the development of semisynthetic or totally synthetic polyamides which would have improved mechanical properties and provide better release kinetics.

Synthetic polyamides offer the additional advantage that the erosion rate can be controlled by the hydrophilicity of the polymer. Thus, in a series of copolymers of L-aspartic acid (hydrophilic) and leucine, β-methyl-L-aspartate, and β-benzyl-L-aspartate (hydrophobic), erosion rates were related to the percentage of L-aspartic acid in the polymer.[13]

Poly(γ-ethyl-L-glutamate-co-L-glutamic acid) (see Figure 2) is synthesized by selective alkaline hydrolysis of poly(γ-ethyl-L-glutamate).[14] The polymer is hydrophilic and water absorption increases with the molar fraction of glutamic acid up to about 50 mol %, beyond which the polymer is water soluble.[15]

The degradation mechanism of the copolymer involved two stages. Initially the ethyl ester side chains were hydrolyzed to form a water-soluble copolymer (about 45 mol% glutamic acid). This copolymer diffused away from the implant in the second step. The degradation rate increased with increasing glutamic acid content. This is probably a combination of increasing water absorption and fewer ethyl ester groups that needed to be hydrolyzed before the copolymer solubilized.

Poly(γ-ethyl-L-glutamic acid) is biocompatible since the dissolved copolymer was absorbed by organs such as the liver and kidney where it was enzymatically degraded to the naturally occurring L-glutamic acid. Ethanol also was a hydrolysate.[15] In order to investigate the suitability of this polyamide to drug delivery applications, cylindrical matrices (0.12 or 0.24-cm diameter) were prepared by extruding a paste consisting of 45% polymer, 45% norgestrel or progesterone (low water solubility), and 10% tetrahydrofuran (THF). The extruded cylinders were dried to remove the THF. Capsules were prepared from tubes and caps were made by dip-coating glass mandrels in polymer solution. An extruded rod of naltrexone (high water solubility) and sesame oil was inserted into the tubes which were capped and sealed with polymer solution. The capsules were 50 μm thick.[15]

Permeabilities were measured at 37°C for various drugs through poly(γ-ethyl-L-glutamate-co-L-glutamic acid) films and the general trend was that drugs with higher water solubilities had higher permeabilities. Permeation rates increased for all drugs as water content (as a result of higher percentages of glutamic acid in the copolymer) of the films increased. The same trend was observed for in vivo naltrexone release from capsules. Release was nonzero order and drug dissolution in the reservoir was rate limiting for high glutamic acid content copolymers. Release from cylindrical matrices was affected by the low permeability of animal tissue to drugs of low water solubility. The passage of drug through the tissue proved to be rate limiting rather than diffusion of drug from the matrix or polymer erosion.[15] The high rate of water absorption of poly(γ-ethyl-L-glutamate-co-L-glutamic acid) precludes its use to achieve a system which degrades by surface erosion.

An interesting abnormality is represented by a 1:1 copolymer of glutamic acid and leucine. When implanted subcutaneously in rats, no weight loss of the implanted device was observed over a period of 15 months, despite the fact that the same material could be completely dissolved by heating it with dilute sodium hydroxide. This abnormality was attributed to structural changes during fabrication of the device, which made the amide linkage inaccessible to proteolytic enzymes.[16,17]

The erosion profile of cross-linked poly(2-hydroxyethyl-L-glutamine) reveals another abnormality. When samples of this polymer were implanted subcutaneously in rats, biodegradation seemingly ceased after only 2 weeks, leaving a significant fraction of the polymeric device uneroded. This phenomenon has been attributed to the increased activity of proteolytic enzymes during the initial acute and chronic stages of inflammation at the implantation site. Thereafter tissue encapsulation of the device seemed to protect the polymer from further enzymatic degradation.[18,19]

Clearly the reliance on proteolytic enzymes as the sole mechanism of degradation for polyamides has several disadvantages: enzymatic activities differ not only from species to species, but also for individuals within the same species. Moreover, as the behavior of poly(2-hydroxyethyl-L-glutamine) indicates, enzymatic activities and hence enzyme-mediated erosion rates might be influenced by intrinsically uncontrollable parameters such as the intensity of the individual inflammatory response at the implantation site. Hence the enzymatic degradation of polyamides does not seem to be a reliable, reproducible, and generally applicable pathway for the formulation of controlled release drug delivery systems.

B. Polyesters

Poly(lactic acid) (PLA), poly(glycolic acid) (PGA), and poly(ε-caprolactones) are the most extensively studied bioerodible polyesters. Several additional polyesters are listed in Table 1. Initially PLA, PGA, and their copolymers were developed as synthetic, absorbable sutures in the 1960s and early 1970s.[20] The success of PLA as a suture and its approval by the Food and Drug Administration (FDA) for human use motivated researchers to investigate the possible application of various polyesters to prostheses[21] and drug delivery systems.[22]

1. Poly(lactic Acid) and Poly(glycolic Acid)

Lactic acid contains a chiral center so there are two enantiomers: D(−)-lactic and L(+)-lactic acid (see Figure 3). Polymers and copolymers of lactic and glycolic acid can be polymerized directly by a polycondensation mechanism. This synthetic method is limited to lower molecular weights so the preferred method is catalytic ring-opening polymerizations of the corresponding cyclic dimers: L(+)-lactide, D(−)-lactide, and glycolide (see Figure 4).[23-26] The *meso*-lactide (Figure 4) is normally not used.[26] Since the polymers are synthesized from the above lactones, they are often called poly(lactide) and poly(glycolide) or poly(dilactide) and poly(diglycolide) in the literature.

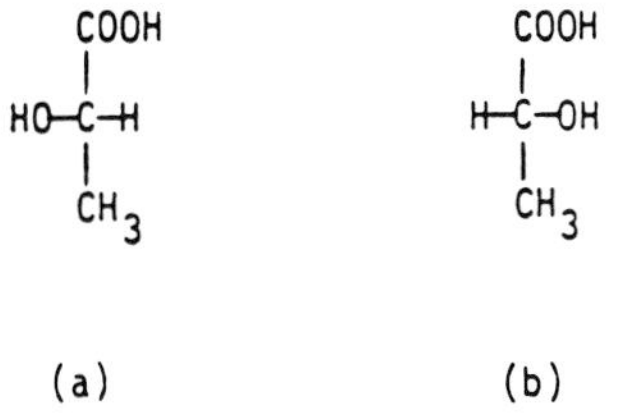

FIGURE 3.   Lactic acid structure: (a) L(+)-lactic acid and (b) D(−)-lactic acid.

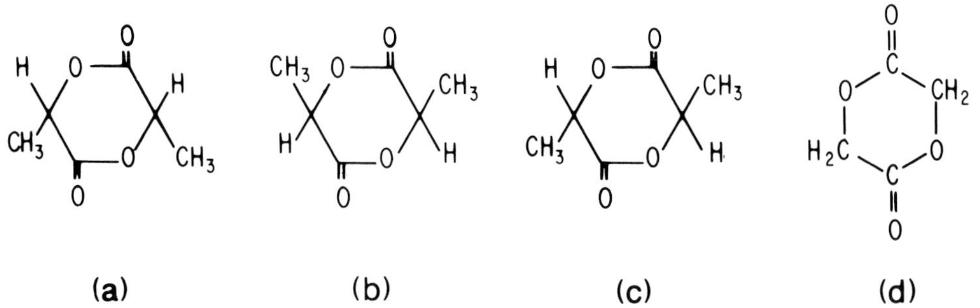

FIGURE 4.   Lactide and glycolide structures: (a) L(+)-lactide, (b) D(−)-lactide, (c) *meso*-lactide, and (d) glycolide.

Poly(D[−]-lactic acid) ([−]PLA) or poly(L[+]-lactic acid) ([+]PLA) are semicrystalline (37%), isotactic, tough, and inelastic with a melting temperature (T_m) of 180°C and a glass transition temperature (T_g) of 57°C. Poly(DL-lactic acid) is completely amorphous, atactic, tough, and inelastic and has a T_g of 57°C. PGA is semicrystalline (46 to 52%), tough, and inelastic with a T_m of 230°C and a T_g of 36°C.[23] Crystallinities, T_gs, and T_ms for copolymers of poly(L[+]-lactic acid-co-glycolic acid), ([+]PLA/PGA), are shown in Figure 5a and b.

The copolymer using PLA instead of (+)PLA is amorphous from 0 to 70 mol% glycolide. Equilibrium water content for (+)PLA/PGA increases with glycolide content until the crystalline region at high mole percent glycolide is reached (see Figure 5c). This trend is expected since the copolymer becomes more hydrophilic as glycolic acid is added.[23]

The in vitro degradation mechanism of PGA (in the form of sutures) has been studied by Chu.[27,28] He proposed a microfibrillar model for the structure of the semicrystalline fibers. Crystalline and amorphous regions alternate along the fiber axis and polymer chains pass between the regions (called tie-chains). The interconnecting tie-chains transmit tensile loads to the crystalline regions. PGA eroded homogeneously, but not at the same rate throughout the whole suture. Chu proposed that hydrolysis began in the amorphous regions where water could most easily penetrate. The backbone ester linkages in the tie-chain fragments which passed through the amorphous regions were hydrolyzed. This reduced the degree of entanglement and allowed undegraded chain segments in the amorphous regions to rearrange into a crystalline structure. The maximum crystallinity measured during degradation was 52%; apparently crystallite hydrolysis (which occurred more slowly) began before the amorphous areas were fully eroded. Hydrolysis of the amorphous regions dominated for the first 21 days. Tensile strength was lost continuously and was completely gone (all tie-chains cleaved) after 49 days. At 60 days, about 50% of the polymer was degraded, yet the gross morphological shape of the suture was still unchanged.[28]

The relative degradation rates in vivo of (+)PLA/PGA copolymer pellets are shown in Figure 6. The graph shows that the half-life of the implanted pellets (5 to 6 mg in rats) quickly dropped from 5 months for PGA to about 1 week for a 50 (+)PLA/50 PGA copolymer

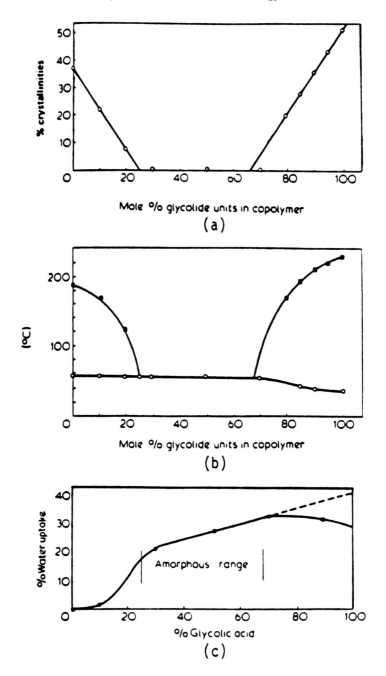

FIGURE 5. Crystallinities, melting points, glass transition temperatures, and water absorption for poly (L [+]-lactic acid-co-glycolic acid): (a) percent crystallinities as a function of composition determined by X-ray and differential scanning calorimetry measurements; (b) melting points and glass transition temperatures measured by differential scanning calorimetry: ●, melting points and ○, glass transition temperatures; and (c) water absorption. (From Gilding, D. K. and Reed, A. M., *Polymer*, 20, 1459, 1979. By permission of the publishers, Butterworth & Co. (Publishers) Ltd.©)

and increased sharply to over 6 months for (+)PLA. The polymer molecular weights were similar (approximately 40,000 to 80,000) and no preferential hydrolysis of glycolide or lactide units in the copolymers was observed.[29] Comparison of Figures 5a and 6 suggests

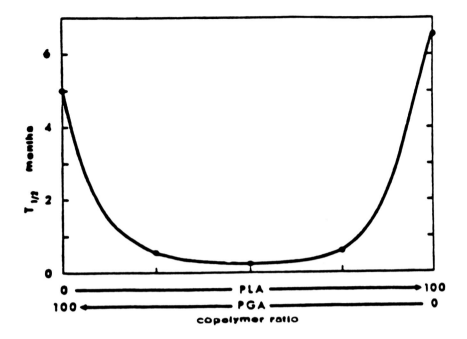

FIGURE 6.   Half-life in vivo for poly(L[+]-lactic acid-co-glycolic acid). Samples implanted in rat tissue. (From Miller, R. A., Brady, J. M., and Cutright, D. E., *J. Biomed. Mater. Res.*, 11, 711, 1977. Reprinted by permission of the copyright owner, John Wiley & Sons,© 1985).

that the erosion rates are a function of polymer crystallinity. An erosion mechanism similar to the one suggested by Chu[28] for PGA probably occurs in the copolymers of (+)PLA/ PGA.

The degradation of poly(DL-lactic acid) also has been studied. Pitt and co-workers[30] observed homogeneous erosion in vivo. Degradation by random chain scission occurred at an even rate throughout the sample since it was uniformly amorphous. Mass loss began at a number average molecular weight of about 15,000 and was accompanied by an increase in hydrolysis rate.[30] Homogeneous erosion at a similar rate has been observed in vitro as well,[24] which implied that the degradation process is nonenzymatic.[30] Mason and co-workers[26] calculated rate constants for the first stage of PLA degradation in various mediums and at different temperatures. PLA erosion rate increased in blood where absorbed lipids may have acted as plasticizers which increased chain mobility. The first-order rate constant for PLA also increased by approximately two orders of magnitude between 25 and 50°C in deionized water.

The erosion of a series of poly(DL-lactide-co-L-lactide)(PLA/[+]PLA) and PLA/PGA polymers has been investigated by Gregory and co-workers,[31] who reported an approximate ranking of erodibility, in the order of most rapid hydrolysis: 75 PLA/25 PGA > 75 (+)PLA/ 25 PGA > 90 PLA/10 PGA > 50 PLA/50 (+)PLA > PLA > (+)PLA.

Lactic acid is a naturally occurring product of glycolysis[32] and labeling studies on PLA and PGA have shown that the polymers are metabolized to carbon dioxide and water.[33] PLA has been classified as minimally toxic.[34] There have been numerous studies and reports on the biocompatibility of these homo- and copolymers.[34-38]

The homo- and copolymers of lactic and glycolic acids were first applied to controlled drug release in the early 1970s.[39,40] Since then, many different device and polymer formulations have been used for bioerodible drug delivery systems (also for agricultural systems)[41,42] and some of these have been reviewed.[43] Several of these formulations and their drug delivery mechanisms are discussed below.

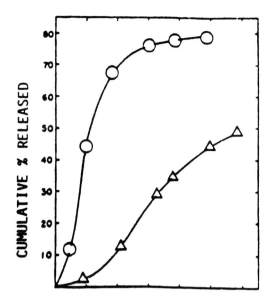

FIGURE 7.   Cyclazocine release from (+)PLA films: ○,
in vivo (implanted in rats) and △, in vitro. (Reprinted with
permission from Woodland, J. H. R., Yolles, S., Blake,
D. A., Helrich, M., and Meyer, F. S., *J. Med. Chem.*, 16,
897, 1973. Copyright 1973 American Chemical Society.)

Yolles and co-workers[44] have formulated films by solvent-casting polymer/drug solutions
and then melt-pressing the dried cast material. In some studies, the films were cryogenically
ground to form polymer/drug particles.[44] Wise and collaborators formulated particles directly
from cast films.[45] Pitt and co-workers[46] formed thin films simply by solvent casting. When
thicker matrices were desired, two or three of the thinner films were stacked and melt-
pressed. Yolles et al.[47] made spherical matrices (beads) by adding an aqueous solution of
dispersing and wetting agents to an organic polymer/drug solution under strong agitation.
Beads were formed by Schwope et al.[36] by molding cast films. They also formulated
cylindrical matrices (rods) by extruding cast films.[36] Matrices have been modified by dip-
coating in pure polymer solutions,[36] sandwiching polymer/drug films between polymer films,[37]
and annealing.[47] Drug delivery devices also have been formulated by microencapsulation[48,49]
and spray drying.[45]

In an additional study, Yolles and co-workers[50] have investigated the release of contraceptive steroids, narcotic antagonists, and anticancer agents from films, beads, and particles
of (+)PLA.[50] In particular, cyclazocine release in vitro and in vivo from (+)PLA (molecular
weight 70,000) films (4 cm² × 0.04 cm) containing 20 wt% drug and 5 wt% tributyl citrate
(a plasticizer) is shown in Figure 7. The unexpected increase in release in vivo was probably
caused by tissue irritation. Lower polymer molecular weight (45,000) did not alter in vivo
release rates significantly even though the polymer was observed to degrade faster.[37] Yolles
et al. had wanted to design a bioerodible device which delivered all the incorporated drug
by a diffusion mechanism followed by polymer erosion.[52] It is likely that the drug delivery
mechanism in Yolles'[50] devices is a combination of diffusion and erosion. The plasticizer
tributyl citrate also complicates the release mechanism as it lowers T_g of (+)PLA which
increases drug permeation rates through the matrix. The tributyl citrate is easily leached
from the polymer; the loss is accompanied by a drop in the drug permeation rate.[53]

Schwope and co-workers[36] have investigated the in vitro and in vivo release of naltrexone
and naltrexone pamoate from 75 (+)PLA/25 PGA, 90 (+)PLA/10 PGA, and (+)PLA.

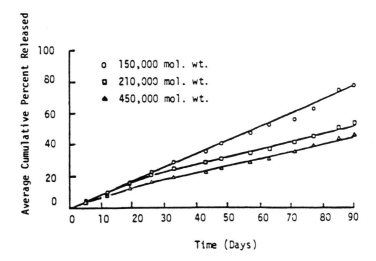

FIGURE 8. Sulfadiazine release from 50 PLA/50 (+)PLA beads. Curves show effect of different molecular weight copolymer on in vitro release rates. (From Wise, D. L., McCormick, G. J., Willet, G. P., Anderson, L. C., and Howes, J. F., *J. Pharm. Pharmacol.*, 30, 686, 1978. With permission.)

They identified many parameters that influence delivery rates. The results of their studies using rods and beads gave a relative naltrexone release rate as a function of polymer composition of 75 (+)PLA/25 PGA > (+)PLA > 90 (+)PLA/10 PGA. Drug solubility in aqueous solutions was directly related to in vivo release rate; the less water-soluble naltrexone pamoate was delivered more slowly than naltrexone. Also at a given water solubility, increased drug solubility in the polymer decreased drug release rate. In rods, the drug loading between 50 and 80 wt% was proportional to the in vitro drug release rate. Dip-coating beads and rods with pure polymer solutions reduced release rates both in vivo and in vitro. Schwope et al.[36] hypothesized that drug release was the result of a combined diffusion and erosion mechanism. A second study (see Figure 8) measured the effect of polymer molecular weight on sulfadiazine release in vitro from 1.5-mm-diameter beads of 50 PLA/50 (+)PLA.[51] A further study correlated in vitro levonorgestrel release to the hydrolytic instability of the polymer used.[31] In an in vivo investigation, Wise and co-workers[54] formulated cylindrical matrices of 90 (+)PLA/10 PGA containing 50 wt % drug. Release of the steroid was maintained for a 2-year period in rats. Release rates fluctuated although there were periods of zero-order release. The recovered rods were brittle, friable, and encapsulated by tissue.[54] Tissue encapsulation may mask the actual behavior of controlled delivery devices in vivo since an aqueous boundary layer forms around the device. Drug transport across the boundary layer becomes the rate-limiting step, resulting in apparent zero-order kinetics.[6]

Pitt and co-workers[46] have studied progesterone release from films of PLA and PLA/PGA. For a thin (3 μm) drug/PLA film sandwiched between two drug-free PLA films (3 μm), ideal diffusion-controlled kinetics was observed in both in vitro and in vivo testing. A thicker, unsandwiched film (100 μm) of PLA with 10 wt% progesterone showed erratic release rates in vitro and in vivo, implying that leaching and/or polymer erosion were complicating drug release. Release duration was 9 weeks in vitro and 15 weeks in vivo. In films (100 μm) of about 80 PLA/20 PGA with 10 wt% progesterone, drug was initially released in vitro by diffusion at a very slow rate followed by a sudden increase in rate (see Figure 9). The sudden increase coincided with the mechanical deterioration and fragmentation of the films which was probably caused by polymer hydrolysis and resulted in the exposure of more surface area.[46]

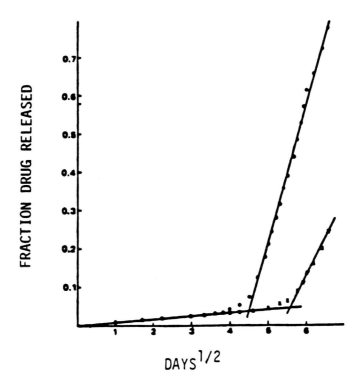

FIGURE 9.   Progesterone release from PLA/PGA films. Release rates were measured in vitro. (From Pitt, C. G., Gratzl, M. M., Jeffcoat, A. R., Zweidinger,
R., and Schindler, A., *J. Pharm. Sci.*, 68, 1534, 1979. With permission.)

Sanders and co-workers[77] have incorporated nafarelin acetate, a potent analogue of luteinizing hormone-releasing hormone, in (+)PLA/PGA microspheres. A triphasic release
of the compound was observed. The first phase was presumed to be compound loss by
diffusion from the surface of the microspheres. The secondary phase occurred concomitantly
with polymer hydrolysis and a decrease in molecular weight, although it remained insoluble.
The third phase involved dissolution of low molecular weight polymer fragments and erosion
of the bulk of the polymer matrix.[77]

2. Polycaprolactone

Poly(ε-caprolactone) and its copolymers with other lactones such as DL-lactide have been
developed by Pitt and co-workers[55] for use in drug delivery devices. Poly(ε-caprolactone)
is synthesized by catalytic ring-opening polymerization[56] (see Figure 10). The polymer is
crystalline (45 to 60%),[53] tough, and flexible. It has a T_m of 63°C and a T_g of −65°C.[55]

A homogeneous erosion mechanism was assumed for the in vivo degradation of poly(ε-
caprolactone) since a tenfold increase in the surface-to-volume ratio did not affect the rate
of degradation.[24] Weight loss began once a limiting molecular weight, M_n, of abut 5000
was reached.[30] Degradation was accompanied by a slow increase in crystallinity, e.g., 45
to 58% in 220 days,[46] as a result of preferential random chain cleavage in the amorphous
regions of the polymer and crystallization of the resulting unrestrained tie segments.[30] The
observation of similar degradation rates in vitro suggests that enzymatic activity is not
responsible for matrix breakdown.[55]

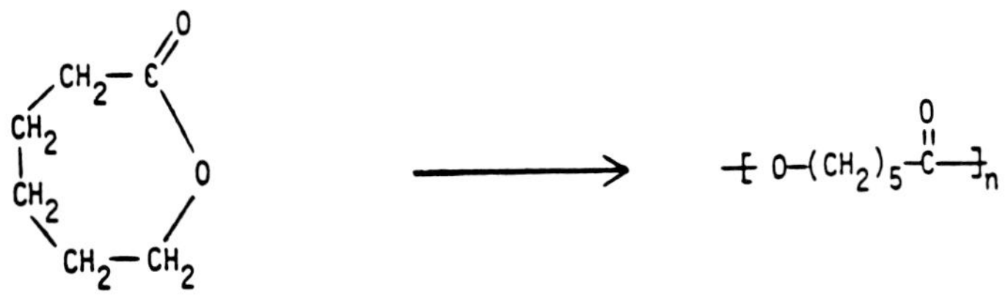

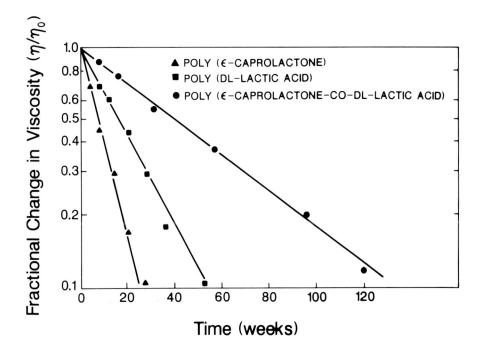

FIGURE 10. Polymerization of ε-caprolactone.

FIGURE 11.   Relative rates of biodegradation of homo- and copolymers of ε-caprolactone and DL-lactic acid. Samples implanted in vivo in rabbits. (From Pitt, C. G., Marks, T. A., and Schindler, A., in *Controlled Release of Bioactive Materials*, Baker, R. W., Ed., Academic Press, New York, 1980, 19. With permission.)

Copolymers of ε-caprolactone and DL-lactide qualitatively showed a similar erosion process. In general, the copolymers degraded more rapidly than the respective homopolymers, apparently due to reductions in crystallinity and lower T_gs. No preferential erosion of one monomer over the other was detected.[30] Degradation rates in vivo for poly(ε-caprolactone), poly(DL-lactic acid), and their copolymers are compared in Figure 11.

Poly(ε-caprolactone) hydrolyzed to ε-hydroxycaproic acid.[30] The polymer demonstrated no adverse tissue reaction in organ and cell cultures (in vitro) or rat muscle tissue (in vivo).[34]

The main attraction of poly(ε-caprolactone) is that the permeation rates of many steroids through the polymer are very high and are similar to the rates through silicone rubber, which has been widely studied for use in inert controlled release devices.[53] For example, poly(ε-caprolactone) is about 10,000 times more permeable to progesterone than poly(DL-lactic acid).[55] Pitt and co-workers[55] have tried to exploit this property to develop a reservoir device which remains intact over a period of about 1 year until the drug is depleted. Copolymers of ε-caprolactone and DL-lactide are suitable for devices with life spans less than 1 year.[24]

Capsules of poly(ε-caprolactone) and its copolymer with DL-lactide were prepared from

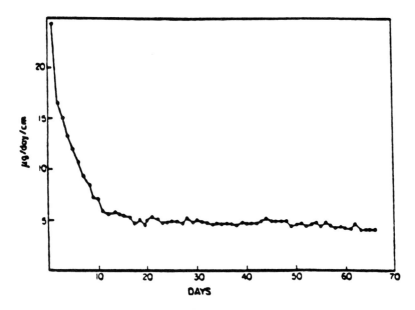

FIGURE 12.   Norgestrel release from poly (ε-caprolactone) capsules. Release rates were measured in vitro; capsule OD/ID = 2.54/1.26. (From Schindler, A., Jeffcoat, R., Kimmel, G. L., Pitt, C. G., Wall, M. E., and Zweidinger, R., *Contemp. Top. Polym. Sci.,* 2, 251, 1977. With permission.)

melt-extruded polymer tubing or from polymer film rolled around a Teflon® core and annealed. The tubing ends were heat sealed with warm pliers. Progesterone, testosterone, or norgestrel were micronized and dispersed in a vehicle (e.g., sesame seed oil) before being placed in the capsules. Capsule thickness was typically about 500 μm.[46]

Release rates at 37°C were predicted from permeabilities of poly(ε-caprolactone) films measured in a diffusion cell. Actual release rates in vitro quickly decreased from the predicted levels over the first 20 days, followed by a slower decline in rate over the remaining 100 to 200 days (Figure 12). The initial decline was due to changes in the dissolution rate of the dispersed drug in the sesame oil. The slower, long-term decrease was due to the increase in crystallinity of poly(ε-caprolactone) as a result of polymer hydrolysis. The increased crystallinity reduced the permeability of the polymer to the steroids.[46]

Pitt and co-workers[46] have also studied in vivo and in vitro progesterone (10, 20, and 30 wt%) release from films (100 to 300 μm in thickness) of poly(ε-caprolactone) and its copolymers with DL-lactide or glycolide. Steroid release was diffusion controlled and completed within 24 hr due to high drug permeability through the polymers. No erosion effects were observed. Drug release was faster from melt-pressed films than solvent-cast films.[46]

Extensive release and degradation studies, particularly by Pitt and co-workers, established beyond doubt that poly(ε-caprolactone) as well as copolymers of ε-caprolactone with lactic acid or glycolic acid erode by a homogeneous erosion mechanism. It was therefore assumed that poly(ε-caprolactone) would not be suitable for the formulation of ideal, surface-eroding matrices. Recently, however, Pitt and co-workers[57] reported that samples of cross-linked polycaprolactone showed surface erosion in vivo. Poly(ε-caprolactone) which has been cross-linked to various degrees with 2,2-bis-(ε-caprolactone-4-yl) propane were elastomeric with little or no crystallinity. Cross-linking therefore seemed to prevent the formation of the densely packed, crystalline regions, leading to greater segmental mobility of the polymer chains. Pitt et al. reasoned that in low T_g polymers the greater segmental chain mobility could allow the ester group to assume the conformation necessary to interact with the active site of esterases. Hence cross-linked poly(ε-caprolactone) would be susceptible to enzymatic degradation at the polymer surface.

Experimental observations seem to confirm this hypothesis. When samples of cross-linked poly(ε-caprolactone) were degraded in vitro (absence of enzymes), only the regular, well-known homogeneous degradation mechanism operated, leading to degradation kinetics virtually identical to the kinetics observed previously with noncross-linked poly(ε-caprolactone). On the other hand, in vivo degradation (in the rabbit and rat) resulted in nearly immediate weight loss via enzymatic surface degradation, superimposed on the regular, nonenzymatic, homogeneous erosion mechanism. It should be noted that the erosion rates varied considerably depending on the animal.[57]

The fact that poly(ε-caprolactone) could be made to erode via a surface erosion mechanism could facilitate the formulation of zero-order release devices based on a poly(ε-caprolactone) matrix. The work of Pitt et al.[57] also suggests that suitable cross-linking of other homogeneously eroding polymers (such as polyamides, PLA, etc.) could possibly improve both their mechanical properties and their release and degradation characteristics.

C. Polyanhydrides

Aromatic polyanhydrides were first synthesized in 1909 by Bucher and Slade[58] and aliphatic polyanhydrides were first prepared in 1932 by Hill and Carothers.[59] Polyanhydrides did not receive much attention again until they were investigated from the middle 1950s to the early 1960s as polyester fiber replacements.[60]

Polyanhydrides were most commonly prepared by a melt-polycondensation method as reported by Hill and Carothers.[59] The dicarboxylic acid monomers are converted to the mixed anhydride with acetic acid by total reflux in acetic anhydride. The isolated prepolymers then undergo melt-polycondensation *in vacuo* under nitrogen sweep.

The major deficiency of polyanhydrides for use as fiber replacements is their hydrolytic instability. Polyanhydrides hydrolyze to form diacid monomers. Aliphatic polyanhydrides such as poly(sebacic acid anhydride) reportedly would become brittle in 24 hr even when kept in a desiccator.[59] Conix[61] later produced more hydrolytically stable polyanhydrides, poly[bis(*p*-carboxyphenoxy) alkane anhydrides], from aromatic diacids which had excellent film- and fiber-forming properties. The hydrophobicity imparted by the phenylene groups in the backbone renders the polyanhydrides much more resistant to hydrolysis. The hydrolytic resistance of these hydrophobic polymers approached that of poly(ethylene terephthalate) in 1*N* NaOH at 25°C (see Figure 13). The hydrolysis rate of poly(sebacic acid anhydride), however, was orders of magnitude faster. The crystalline polymers were found to be twice as hydrolytically stable as the amorphous ones.

It is the hydrolytic instability of the anhydride linkage which renders polyanhydrides attractive as biodegradable drug-carrier matrices. The water-labile anhydride linkage provides the basis for using a variety of backbones and yet ensuring biodegradability. In one study,[62] a 1-mm-thick slab of poly[bis(*p*-carboxyphenoxy) methane anhydride] (PCPM) was found to hydrolyze completely in 50 days in pH 7.4 phosphate buffer at 37°C, leaving no insoluble residue. The erosion of compression-molded PCPM slabs was characterized by an induction period followed by a period of mass loss at a nearly constant rate (see Figure 14). Throughout the erosion, the devices decreased in size while maintaining their physical integrity, suggesting that surface erosion was occurring. Release of cholic acid from these slabs closely corresponded to the erosion of the matrix. The erosion and release profiles were nearly zero order and had similar slopes (see Figure 15). Both the drug and the polymer completely disappeared at nearly identical times.

Recently, a more complete study was performed on the homologous polymers based on bis (*p*-carboxyphenoxy) propane (PCPP) and hexane.[63] As the alkane in the backbone was changed from a methyl to a hexyl group, the erosion rates underwent a decrease of three

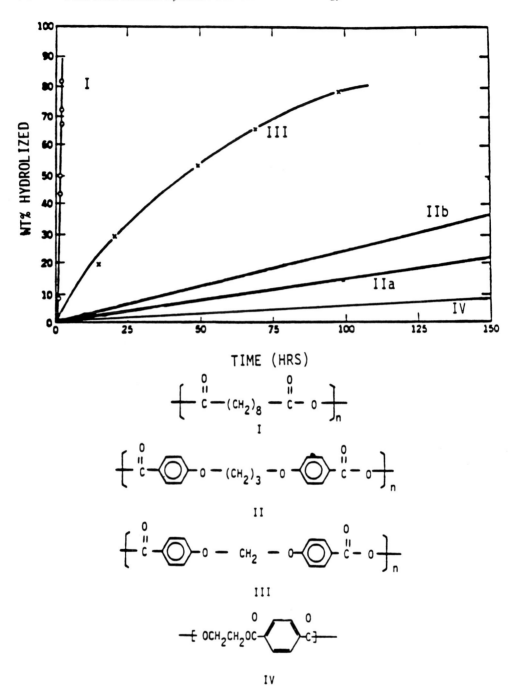

FIGURE 13. Stability against hydrolysis of various polyanhydrides and polyesters (1*N* sodium hydroxide): (I) poly(sebacic acid); (IIa) crystalline polyanhydride from 1,3-PCPP; (IIb) amorphous polyanhydride from 1,3-PCPP; (III) polyanhydride from bis(*p*-carboxyphenoxy)methane; (IV) polyester from ethylene terephthalate. (From Conix, A., *J. Polym. Sci.*, 29, 343, 1958. Reprinted by permission of the copyright owner, John Wiley & Sons, © 1985.)

orders of magnitude. The degradation rates of these polymers could be enhanced by incorporating a more hydrophilic comonomer, sebacic acid (SA) (see Figure 16). The more hydrophobic polymers, PCPP and PCPP-SA 85:15, displayed constant erosion kinetics over several months. By extrapolation, PCPP will completely degrade in over 3 years. An increase

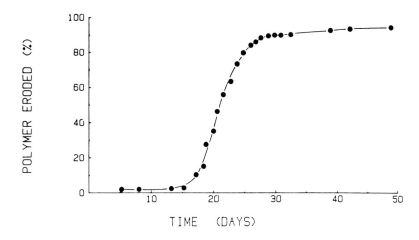

FIGURE 14.   Erosion curves for drug-free PCPM in 0.1 *m* phosphate buffer, pH 7.4, at 37°C. (From Rosen, H. B., Chang, J., Wnek, G. E., Linhardt, R. J., and Langer, R., *Biomaterials*, 4, 131, 1983. With permission.)

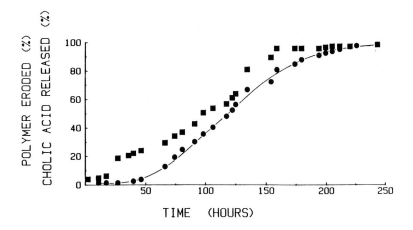

FIGURE 15.   Erosion and release curves for a PCPM matrix containing cholic acid at 10.5 wt% eroded in 0.1 *M* phosphate buffer, pH 7.4, at 60°C; ●, polymer erosion and ■ drug release. (From Rosen, H. B., Chang, J., Wnek, G. E., Linhardt, R. J., and Langer, R., *Biomaterials,* 4, 131, 1983. With permission.)

of 800 times in erosion rate was observed when the SA content reached 80%. These more hydrophilic copolymers tended to crumble toward the later stages of erosion. Erosion of these polyanhydrides was pH sensitive, enhanced in high pH, and became more stable in acidic conditions (see Figure 17).

In release studies, the drug release profile of a model drug *p*-nitroaniline followed closely that of the degradation of injection-molded PCPP over a period of more than 8 months, suggesting a release mechanism that was dominantly erosion controlled (see Figure 18). However, this was true only for injection-molded samples. Release from compression-molded devices was no longer concomitant with polymer degradation.[63]

In addition to determining the erosion and release characteristics of these polyanhydrides; the chemical reactivity and biocompatibility of these polyanhydrides were assessed.[64] By design the polymer is meant to react with water, thereby leading to erosion and drug release. However, this reactivity also raises the concern of the reaction of nucleophiles other than

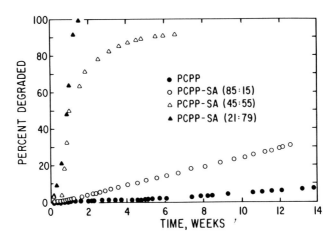

FIGURE 16. Degradation profiles of compression-molded poly[bis(*p*-carboxyphenoxy)propane anhydride] and its copolymers with SA in 0.1 *M* pH 7.4, phosphate buffer at 37°C. (From Leong, K. W., Brott, B. C., and Langer, R., *J. Biomed. Mater. Res.,* 19, 941, 1985. Reprinted by permission of the copyright owner, John Wiley & Sons, © 1985.)

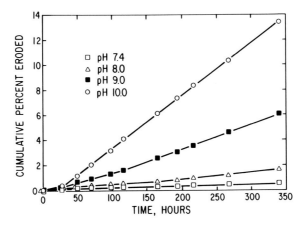

FIGURE 17. pH dependence of erosion rates of compression-molded poly[bis(*p*-carboxyphenoxy)propane anhydride] in 0.1 *M* phosphate buffers at 37°C. (From Leong, K. W., Brott, B. C., and Langer, R., *J. Biomed. Mater. Res.,* 19, 941, 1985. Reprinted by permission of the copyright owner, John Wiley & Sons, © 1985.)

water with the anhydride linkage. Using the amino group as a model functionality, it was observed by infrared spectroscopy that PCPP-SA 21:79 reacted with *p*-phenylenediamine, forming amides when the sample was injection-molded at 120°C. It appeared the drug-matrix interaction was thermally triggered as no reaction was observed when the samples were compression-molded at room temperature. It was also observed that no reaction occurred between the polymer and the model drug *p*-anisidine during the hydrolytic degradation of the matrix at 37°C.

These polyanhydrides showed favorable biocompatibility from tissue response and toxicological studies. The polymers did not provoke inflammatory responses in the corneas of rabbits over a 6-week implantation period. In comparison to other previously tested polymers, the inertness of these polyanhydrides rivals that of the biocompatible poly(hydroxyethyl

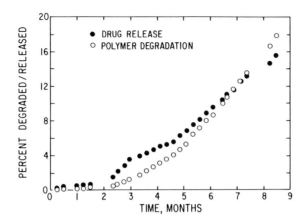

FIGURE 18.   Release of *p*-nitroaniline (10% loading) from injection-molded poly[bis(*p*-carboxyphenoxy)propane anhydride] in 0.1 *M*, pH 7.4, phosphate buffer at 37°C. (From Leong, K. W., Brott, B. C., and Langer, R., *J. Biomed. Mater. Res.*, 19, 941, 1985. Reprinted by permission of the copyright owner, John Wiley & Sons, © 1985.)

methacrylate) and poly(ethylene vinyl acetate).[65] The host response to polymers implanted subcutaneously in rats over a 6-month period was also mild. In toxicological studies, neither mutagenicity nor toxicity was associated with the polymers or their breakdown products. The products also gave a negative response in an in vitro teratogenicity assay. In addition, the in vitro growth of mammalian cells on the polymers was unaffected as measured by cell morphology and cell growth rate.[64] These polymers containing the nitrosourea BCNU have recently been approved by the U.S. Food and Drug Administration for clinical tests on human brain cancer. At this writing, five patients have been treated with this polymer-drug delivery system.

D. Poly(orthoesters)

Poly(orthoesters) can be prepared by a transesterification reaction:[66]

$$\underset{O}{\overset{EtO\quad OEt}{\times}} + HO-R-OH \rightarrow \underset{O}{\overset{[O\quad O-R]_{n+EtOH}}{\times}}$$

and also by reaction of diols with diketene acetals:[67]

$$CH_2{=}C{-}O{-}R'{-}O{-}C{=}CH_2 \;(OR) + HO{-}R''{-}OH \rightarrow \left[O{-}\underset{CH_3}{\overset{OR}{C}}{-}O{-}R'{-}O{-}\underset{CH_3}{\overset{OR}{C}}{-}O{-}R'' \right]_n$$

Reaction of the diols and diketene acetals is extremely rapid. The addition polymerization can be conducted at atmospheric pressure without external heating since the reaction is exothermic. Hydrolysis of orthoesters is acid catalyzed; the orthoester bond is relatively stable in alkaline conditions. Initial attempts to develop a surface eroding system incorporated basic salts along with drugs into the matrix to deter biodegradation in the bulk.[68]

A poly(orthoester) system developed by Alza and identified as Chronomer™ (now called

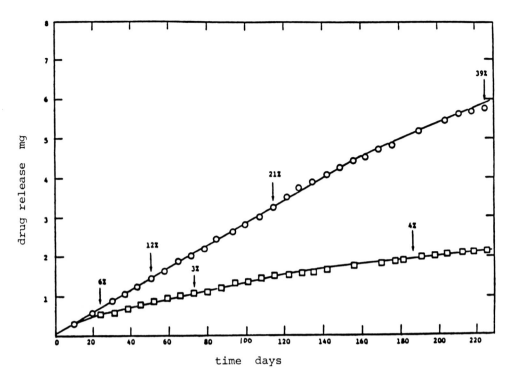

FIGURE 19.   Norethindrone release from a 3,9-bis(methylene)-2,4,8,10-tetraoxaspire[5,5]undecane/1,6-hexane-
diol poly(orthoester). Polymer was loaded with 10 wt% sodium carbonate and varying amounts of norethindrone
at pH 7.4 and 37°C (arrows indicate percent weight loss); ○, 20 wt% (8.0 mg) drug and (□ 10 wt% (4.0 mg) drug.
(From Heller, J. Penhale, D. W. H., Helwing, R. F., and Fritzinger, B. K., *Polym. Eng. Sci.*, 21, 727, 1981.
With permission.)

Alzamer™) was used for releasing naltrexone and contraceptive steroids. Polymers with
properties ranging from tough and glassy to soft and compliant could be prepared by changing
the chemical structure of the diols.[69] These hydrophobic polymers reportedly eroded heterogeneously with induction periods observed in both in vivo and in vitro conditions.[69] In
vitro release of naltrexone from disks of soft and complaint Chronomer™ containing 10 wt
% sodium carbonate and up to 30 wt% drug was studied.[69] The basic salt, sodium carbonate,
was added to prevent degradation in the bulk. Zero-order release kinetics was observed with
complete polymer erosion coinciding with drug exhaustion. Similar devices made from glassy
and tough Chronomer™ showed that erosion lagged behind drug release. Both systems
increased in size and weight due to water absorption after implantation during in vivo studies.
Interestingly, the in vivo release from cylindrical devices was determined to be zero order.
Theoretically, drug release from a heterogeneously eroding cylinder should decrease with
time as surface area diminishes. Benagiano et al.[70] hypothesized that the erosion was au-
tocatalyzed by acidic erosion products.
    Another poly(orthoester) system also developed by Heller was prepared by addition of
1,6-hexanediol or *trans*-cyclohexane-dimethanol to 3,9-bis[ethylidene-2,4,8,10-tetraoxas-
piro (5,5)-undecane.] A disk fabricated from this polymer showed a 2-month induction
period. This was followed by accelerated erosion due to an increase in polymer hydrophilicity
caused by the water-soluble erosion products. In vitro release from disk-shaped devices
containing 10 wt% norethindrone and 10 wt% sodium carbonate was constant for more than
8 months (see Figure 19). The weight loss data, however, showed that the release was not
controlled by polymer erosion.[68,71] A swelling-controlled mechanism was hypothesized to
explain the linear release. It was suggested that water diffused into the matrix to dissolve

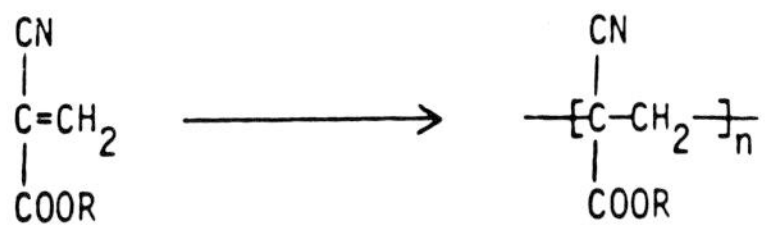

$$\text{CN} \qquad\qquad \text{CN}$$
$$| \qquad\qquad\qquad |$$
$$\text{C=CH}_2 \longrightarrow \text{-}[\text{C-CH}_2\text{-}]_n$$
$$| \qquad\qquad\qquad |$$
$$\text{COOR} \qquad\qquad \text{COOR}$$

FIGURE 20. Polymerization of alkyl 2-cyanoacrylates.

the sodium carbonate, which in turn exerted an osmotic pressure on the polymer. The polymer swelled and the drug was released due to the relaxation of the matrix. The constant release rate was attributed to the constant progression of a swelling front analogous to Case II diffusion.[72,73] Replacing the basic salt by a neutral osmotically active salt such as sodium sulfate yielded constant release in the initial stage (60 days), but the release rate increased significantly afterwards.[74] In this case, degradation in the bulk was not impeded, and as a result the cell walls ruptured, yielding a foam-like structure which promoted diffusion through pores.

Recent approaches taken by researchers to achieve zero-order erosion for poly(orthoesters) were to catalyze hydrolysis at the matrix surface rather than suppressing degradation in the bulk[75,76] Using calcium lactate (2 wt%) as a catalyst, Heller[75] was able to obtain constant release of levonorgestrel for more than a year from cylindrical devices containing 30 wt% of drug. Correlation between polymer erosion and drug release was close. In vivo studies in rabbits showed that comparable erosion and release characteristics were obtained over 4 months. Sparer et al.,[76] using the more acidic catalysts, acid anhydrides, were able to obtain much higher erosion rates. Zero-order and concomitant erosion and release were achieved by a surface reaction zone mechanism, although only a low loading level of 0.2% was used. According to this mechanism, zero-order release can be obtained only if hydrolysis of orthoester linkages or acid anhydride catalysts, or both, is rate limiting. The release was affected by a number of factors. Increasing the acid strength, drug loading, and amount of catalysts (up to 2 wt%) would markedly increase the release rate. Increasing the molecular weight and glass transition temperature would, on the other hand, decrease the release rate.

E. Poly(alkyl 2-Cyanoacrylates)

Alkyl 2-cyanoacrylate monomers[78] and their polymers[79] have been used or investigated as surgical adhesives. Poly(n-butyl 2-cyanoacrylate) has been used to microencapsulate proteins for controlled release. Poly(alkyl 2-cyanoacrylates) (see Figure 20) can be made by bulk, anionic,[80] or interfacial polymerization.[81]

Vezin and Florence investigated the erosion rates of films and particles of the ethyl, n-butyl, and n-hexyl poly(alkyl 2-cyanoacrylate) homologues. They found that in vitro erosion rate decreased with increasing length of the alkyl group.[80] The same trend was observed in organ culture studies.[35] Vezin and Florence also found that erosion rates were inversely related to molecular weight and directly related to pH. A heterogeneous erosion process for films and particles ($m_n > 1000$) was proposed based on other chemical measurements and electron microscopy. Particle surfaces became pitted as erosion occurred, resulting in an increase in specific area. This resulted in a constant erosion rate (up to about 34% total erosion) rather than the expected decreasing rate. Within the molecular weight range studied, $1000 < M_n\ 10,000$, erosion rates were directly proportional to $1/m_n$, implying that heterogenous polymer erosion occurred by chain-end hydrolysis. Lower molecular weight poly(ethyl 2-cyanoacrylates) eroded homogeneously.[80]

Poly(alkyl 2-cyanoacrylates) degraded to formaldehyde and the corresponding alkyl cyanoacetate.[81] The toxicity of the polymers in vivo was related to some degree to the length of the alkyl substituent; poly(methyl 2-cyanoacrylate) was classified as highly toxic, poly(isobutyl 2-cyanoacrylate) was classified as moderately toxic, and homologues above butyl were listed as minimally toxic.[34] Hegyeli[35] found in organ culture studies that the erosion products from methyl and butyl homologues were equally damaging to tissues. He suggested that the observed trend in toxicity in vivo was caused by differing degradation rates; the quickly degrading poly(methyl 2-cyanoacrylate) resulted in higher concentrations of the toxic erosion products than the less degradable homologues.

Wood and co-workers[81] used poly(*n*-butyl 2-cyanoacrylate) to microencapsulate proteins for controlled release. Microcapsules (25 to 250 μm) were formed by interfacial polymerization of the *n*-butyl 2-cyanoacrylate in water-oil emulsions where albumin or fibrinogen was included in the aqueous phase. After washing, the microcapsules contained about 39 and 65%, respectively, of the albumin or fibrinogen initially used. Ultrasonification of the capsules resulted in only minor losses of protein, implying that it was chemically or physically attached to the polymer surface. Protein which was in the free volume of the microcapsules was probably lost during washing. The in vivo and in vitro behavior of these microcapsules is being investigated.[81]

III. MATHEMATICAL MODELING FOR BIOERODIBLE SYSTEMS

Mathematical modeling of release from bioerodible systems is not as advanced as that of diffusion and swelling-controlled systems because both drug diffusion and matrix erosion have to be considered. Solutions had been obtained by Hopfenberg[82] and Cooney[83] for simplified situations where diffusional contribution to release was neglected.

For an idealized bioerodible delivery system, the drug is released only when the matrix heterogeneously erodes. For a given matrix, the erosion rate is a function only of surface area.[82] Hopfenberg[82] modeled the release from cylindrical, spherical, and flat matrices for these idealized cases. A single relationship was found to describe all three shapes:

$$M_t/M_\infty = 1 - [1 - K_o t/C_o a]^n \tag{1}$$

where M_t is the amount of drug released from the device in time t, M_∞ is the total amount of drug released when the device is exhausted, K_o is the erosion rate constant, C_o is the uniform initial concentration of drug in the matrix, a is the radius for the sphere and cylinder or the half-thickness for a slab, and n is a shape factor with the value of 3 for a sphere, 2 for a cylinder, and 1 for a slab. The model ignores edge and end effects. From Equation 1, the geometry which gives zero-order drug release is the slab. Both the sphere and cylinder result in release rates decreasing with time.

Cooney[83] performed a more detailed analysis of spheres and cylinders undergoing heterogeneous degradation. As the ratio of initial cylinder length to diameter (L_o/D_o) approached zero and the geometry became a flat disk, the release rate approached zero order. The release rates for nonzero L_o/D_o ratios are shown in Figure 21. Cooney also modeled spheres and cylinders with bores. Solution of the equations for cylinders with a concentric bore and coated ends showed zero-order kinetics. Some novel matrix shapes for achieving zero-order release were also proposed.

Baker and Lonsdale[84] introduced an analysis which allowed for diffusion, but assumed a linear concentration profile in the matrix. Using a pseudosteady-state approach, the relationship between the release rate and erosion rate was presented without the analytical solution. By considering both homogeneous and heterogeneous erosion, they derived an expression for the overall erosion rate for a slab geometry:

$$\frac{dx}{dt} = \frac{(K_s - K_b)m}{\ln(A^o/A'')} - tK_b \tag{2}$$

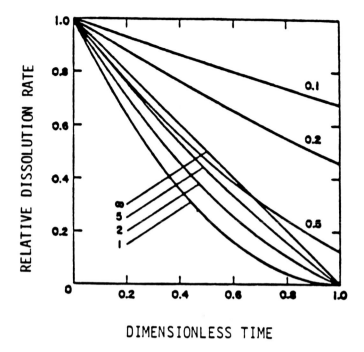

FIGURE 21. Dissolution behaviors of cylinders of various length/diameter ratios. $L_o/D_o < 1$ corresponds to disk-like cylinders and $L_o/D_o < 1$ corresponds to rod-like cylinders. The curve for $L_o/D_o \to 0$ is a horizontal line at a relative dissolution rate of 1.0. (From Cooney, D. O., *AIChE J.*, 18, 446, 1972. Reproduced by permission of the American Institute of Chemical Engineers.)

where K_s and K_b are the surface and bulk reaction rate constants, respectively, m is the thickness of the skin in which the surface reaction is occurring, A_o is the initial concentration of reactable bonds, and A'' is the concentration of these bonds at time t.

Heller and Baker[85] attempted to model release from homogeneously degrading poly(lactic acid) systems, where chain cleavage is first order and diffusion through the matrix is the rate-determining step. By modifying the Higuchi model,[86] an expression for release rate was derived for a film:

$$d\,Mt/dt = \frac{A}{2}\left[\frac{2P_o\,\exp\,(Kt)C_o}{t}\right]^{1/2} \qquad (3)$$

where P_o is the initial permeability of the polymer to the drug and is erosion dependent, A is the total surface area, and K is a first-order rate constant. Release profiles resulting from Equation 3 and pure diffusion are shown in Figure 22. The polymer permeability increases as a result of erosion and compensates for the normal decline in release rate. Although the theory qualitatively matches the experimental release profile, one limitation of the model is that polymer chains in PLA films are cleaved at first-order kinetics only up to a limiting molecular weight of 5000[30] rather than for the whole erosion process as assumed by Heller and Baker.[85]

Lee[87] later developed a more general model which took into account diffusional release as well as different ratios of drug loading to drug solubility in the matrix. In addition to an eroding polymer front, a moving diffusion front which separates an undissolved core with uniformly dispersed solutes from a partially extracted region with dissolved solutes was considered. An approximate analytical solution was provided for conditions of planar geometry, constant surface erosion, perfect sink, and negligible edge effects:

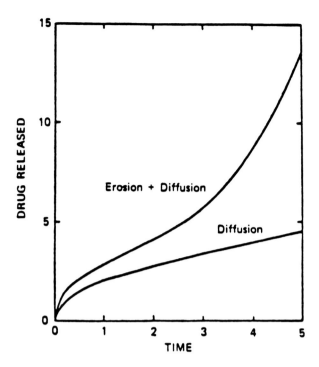

FIGURE 22.   Theoretical curves of drug release for a polymer slab by diffusion and diffusion with erosion. (From Heller, J. and Baker, R. W., in *Controlled Release of Bioactive Materials*, Baker, R. W., Ed., Academic Press, New York, 1980, 1. With permission.)

$$\frac{M_t}{M_\infty} = \delta\left[1 + \left(\frac{K_o a}{D}\right)\tau - \left(\frac{C_s}{C_o}\right)\left(\frac{3 + a_3}{6}\right)\right] \qquad (4)$$

where δ is the dimensionless moving front.

$$\tau = \frac{Dt^2}{a} = \frac{1}{12h}\left[6\left(\frac{C_o}{C_s}\right) - 4 - a_3\right][\delta - {}^1/_2\,h \times \ln(1 + 2\delta h)]$$

$$h = {}^1/_2\left(1 - \frac{C_o}{C_s}\right)\left(\frac{K_o a}{D}\right)$$

$$a_3 = \frac{C_o}{C_s} + \delta h - \left[\left(\frac{C_o}{C_s} + \delta h\right)^2 - (1 + 2\delta h)\right]^{1/2}$$

and a is the half thickness, D is the solute diffusion coefficient in the matrix, C_s is the equilibrium solute solubility in the matrix, R is the position of the moving diffusion front, and S is the position of the moving eroding front. The model predicts that the release would approach zero order if the drug loading is much greater than the drug solubility in the matrix (Figure 23).

Recently Thombre and Himmelstein[88] modeled the release from erodible devices which are laminated with a secondary membrane. The membrane is assumed to be nonerodible during the release process. By considering the release from diffusion to be comparable to that from erosion, the rate equations for the movements of the diffusion and erosion fronts were derived. These equations were solved numerically to obtain cumulative release profiles.

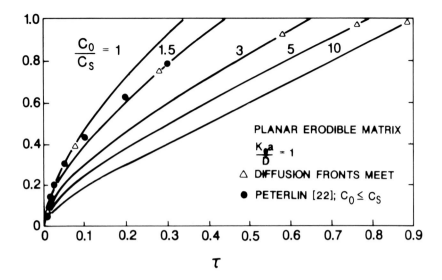

FIGURE 23. Fractional release vs. time for a dispersed solute in a planar erodible polymer matrix (k_o a/D = 1) with various solute loading levels. (From Lee, P. I., *J. Membr. Sci.*, 7, 255, 1980. With permission.)

The analysis shows that the barrier membrane can be used to adjust the duration of the diffusive release of the drug.

ACKNOWLEDGMENTS

This work was supported by the National Institutes of Health (NIH) Grant 26698.

REFERENCES

1. **Baker, R. W. and Lonsdale, H. K.,** Controlled release: mechanisms and rates, in *Advances in Experimental Medicine and Biology,* Vol. 47, Tanquary, A. C. and Lacey, R. E., Eds., Plenum Press, New York, 1974, 15.
2. **Heller, J.,** Controlled release of biologically active compounds from bioerodible polymers, *Biomaterials,* 1, 51, 1980.
3. **Kronenthal, R. L.,** Biodegradable polymers in medicine and surgery, *Polym. Sci. Technol.,* 8, 119, 1974.
4. **Kojima, K.,** Interaction between polymeric materials and tissue, *Bull. Tokyo Med. Dent. Univ.,* 22, 263, 1975.
5. **Langer, R. S. and Peppas, N. A.,** Present and future applications of biomaterials in controlled drug delivery systems, *Biomaterials,* 2, 201, 1981.
6. **Benagiano, G. and Gabelnick, H. L.,** Biodegradable systems for the sustained release of fertility-regulating agents, *J. Steroid Biochem.,* 11, 449, 1979.
7. **Rudakova, T. E., Zaikov, G. E., Voronkova, O. S., Daurova, T. T., and Degtyareva, S. M.,** The kinetic specificity of polyethylene terephthalate degradation in the living body, *J. Polym. Sci. Polym. Symp.,* 66, 277, 1979.
8. **Gumargalieva, K. Z., Moiseev, Y. V., Daurova, T. T., Voronkova, O. S., and Rozanova, I. B.,** Polycaproamide degradation in rabbits and in several model media, *Biomaterials,* 1, 214, 1980.
9. **Moiseev, Y. V., Daurova, T. T., Voronkova, O. S., Gumargalieva, K. Z., and Privalova, L. G.,** The specificity of polymer degradation in the living body, *J. Polym. Sci. Polym. Symp.* 66, 269, 1979.
10. **Rubin, A. L., Stenzel, K. H., Miyata, T., White, H. J., and Dunn, M.,** Collagen as a vehicle for drug delivery, *J. Clin. Pharmacol.,* 13, 309, 1973.
11. **Bradley, W. G. and Wilkes, G. L.,** Some mechanical property considerations of reconstituted collagen for drug release supports, *Biomater. Med. Devices Artif. Organs,* 5, 159, 1977.
12. **Lee, T. K., Sokoloski, T. D., and Royer, G. P.,** Serum albumin beads: an injectable, biodegradable system for the sustained release of drugs, *Science,* 213, 233, 1981.

13. **Marck, K. W., Wildevuur, Ch. R. H., Sederel, W. L., Bantjes, A., and Feijen, J.,** Biodegradability and tissue reaction of random copolymers of L-leucine, L-aspartic acid, and L-aspartic acid esters, *J. Biomed. Mater. Res.,* 11, 405, 1977.

14. **Sidman, K. R., Schwope, A. D., Steber, W. D., and Poulin, S. B.,** Development of Evaluation of a Biodegradable Drug Delivery System, Contract No. N01-HD-4-2802, Contractor Quarterly Rep. for the Period October 1979 to December 1979, National Institute of Child Health and Human Development, Bethesda, Md., 6.

15. **Sidman, K. R., Schwope, A. D., Steber, W. D., Rudolph, S. E., and Poulin, S. B.,** Biodegradable, implantable, sustained release systems based on glutamic acid copolymers, *J. Membr. Sci.,* 7, 277, 1980.

16. **Sidman, K. R., Schwope, A. D., Steber, W. D., and Rudolph, S. E.,** Use of synthetic polypeptides in the preparation of biodegradable delivery system of narcotic antagonists, in Narcotic Antagonists, NIDA Res. Monogr. Ser. No. 28, Willette, R. E. and Barnett, G., Eds., U.S. Government Printing Office, Washington, D.C. 1981.

17. **Sidman, K. R., Steber, W. D., and Burg, A. W.,** A biodegradable drug delivery system, in Proc. Drug. Delivery Systems, DHEW Publ. No. 77-1238, Gabelnick, H. L., Ed., National Institutes of Health, Washington, D.C., 1976.

18. **Dickinson, H. R. and Hiltner, A.,** Biodegradation of poly(α-amino acid) hydrogel. II. In vitro, *J. Biomed. Mater. Res.,* 15, 591, 1981.

19. **Dickinson, H. R., Hiltner, A., Gibbons, D. F., and Anderson, J. M.,** Biodegradation of poly (α-amino acid) hydrogel. I. In vivo, *J. Biomed. Mater. Res.,* 15, 577, 1981.

20. **Laufman, H. and Rubel, T.,** Synthetic absorbable sutures, *Surg. Gynecol. Obstet.,* 145, 597, 1977.

21. **Reid, R. L., Cutright, D. E., and Garrian, S.,** Biodegradable cuff on adjunct to peripheral nerve repair, *Hand,* 10, 259, 1978.

22. **Wise, D. L., Fellmann, T. D., Sanderson, J. E., and Wentworth, R. L.,** Lactic/glycolic acid polymers, in *Drug Carriers in Biology and Medicine,* Gregoriadis, G., Ed., Academic Press, London, 1979, 237.

23. **Gilding, D. K. and Reed, A. M.,** Biodegradable polymers for use in surgery—polyglycolic/poly(lactic acid) homo- and copolymers, *Polymer,* 20, 1459, 1979.

24. **Schindler, A., Jeffcoat, R., Kimmel, G. L., Pitt, C. G., Wall, M. E., and Zweidinger, R.,** Biodegradable polymers for sustained drug delivery, *Contemp. Top. Polym. Sci.,* 2, 251, 1977.

25. **Chujo, K., Kobayashi, H., Suzuki, J., Toluhara, S., and Tanabe, M.,** Ring-opening polymerization of glycolide, *Makromol. Chem.,* 100, 262, 1967.

26. **Mason, N. S., Miles, C. S., and Sparks, H. E.,** Hydrolytic degradation of poly DL-(lactide), *Polym. Sci. Technol.,* 14, 279, 1980.

27. **Chu, C. C.,** An in vitro study of the effect of buffer on the degradation of poly(glycolic acid) sutures, *J. Biomed. Mater. Res.,* 15, 19, 1981.

28. **Chu, C. C.,** Hydrolytic degradation of polyglycolic acid: tensile strength and crystallinity study, *J. Appl. Polym. Sci.,* 26, 1727, 1981.

29. **Miller, R. A., Brady, J. M., and Cutright, D. E.,** Degradation rates of oral resorbable implants (polylactates and polyglycolates): rate modification with changes in PLA/PGA copolymer ratios, *J. Biomed. Mater. Res.,* 11, 711, 1977.

30. **Pitt, C. G., Gratzl, M. M., Kummel, G. L., Surles, J., and Schindler, A.,** Aliphatic polyesters, II. The degradation of poly(DL-lactide), poly(ϵ-caprolactone), and their copolymers *in vivo, Biomaterials,* 2, 215, 1981.

31. **Gregory, J. B., Newberne, P. M., Wise, D. L., Bartholow, L. C., and Stanburg, J. B.,** Results on biodegradable cylindrical subdermal implants for fertility control, in *Polymeric Delivery Systems,* Midland Macromolecular Institute Monogr. 5, Kostelnik, R. J., Ed., Gordon and Breach, New York, 1978.

32. **Lehninger, A. L.,** *Biochemistry,* 2nd ed., Worth, New York, 1975.

33. **Morgan, M. N.,** New synthetic absorbable suture material, *Br. Med. J.,* 2, 308, 1969.

34. **Wade, C. W. R., Hegyeli, A. F., and Kulkarni, R. K.,** Standard for *in vitro* and *in vivo* comparison and qualification of bioabsorbable polymers, *J. Test. Eval.,* 5, 397, 1977.

35. **Hegyeli, A. F.,** Use of organ cultures to evaluate biodegradation of polymer implant materials, *J. Biomed. Mater. Res.,* 7, 205, 1973.

36. **Schwope, A. D., Wise, D. L., and Howes, J. F.,** Lactic/glycolic acid polymers as narcotic antagonist delivery systems, *Life Sci.,* 17, 1877, 1975.

37. **Woodland, J. H. R., Yolles, S., Blake, D. A., Helrich, M., and Meyer, F. J.,** Long-acting delivery systems for narcotic antagonist, *J. Med. Chem.,* 16, 897, 1973.

38. **Blomstedt, B. and Jacobson, S-I.,** Experiences with polyglactin 910 (Vicryl®) in general surgery, *Acta Chir. Scand.,* 143, 259, 1977.

39. **Blake, D. A., Yolles, S., Helrich, M., Cascorbi, H. F., and Eagan, M. J.,** Release of Cyclozone from Subcutaneously Implanted Polymeric Matrices, Abstract Academy of Pharmaceutical Science, San Francisco, 1971.

40. **Wise, D. L.**, Development of a Sustained Release System for Treatment of Narcotic Patients, in *Reports on Dynatech Contract with the City of New York, Dynatech, Cambridge, Mass.*, 1970.

41. **Sinclair, R. G.**, Slow-release pesticide-system polymers of lactic and glycolic acids as ecologically beneficial, cost-effective encapsulating materials, *Environ. Sci. Technol.*, 7, 955, 1973.

42. **Jaffe, H., Miller, J. A., Giang, P. A., and Hayes, D. K.**, Implantable systems for delivery of insect growth regulators to livestock, in *Controlled Release of Bioactive Materials*, Baker, R. W., Ed., Academic Press, New York, 1980, 237.

43. **Yolles, S. and Sartori, M. F.**, Degradable polymers for sustained drug release, in *Drug Delivery Systems*, Juliano, R. L., Ed., Oxford University Press, New York, 1980, 84.

44. **Yolles, S., Leafe, T., Ward, L., and Boettner, F.**, Controlled release of biologically active agents, *Am. Chem. Soc. Div. Org. Coat. Plast. Chem. Pap.*, 36, 332, 1976.

45. **Wise, D. L., McCormick, G. F., Willet, G. P., and Anderson, L. C.**, Sustained release of an antimalarial drug using a copolymer of glycolic/lactic acid, *Life Sci.*, 19, 867, 1976.

46. **Pitt, C. G., Gratzl, M. M., Jeffcoat, A. R., Zweidinger, R., and Schindler, A.**, Sustained drug delivery systems. II. Factors affecting release rates from poly (ϵ-caprolactone) and releated biodegradable polyesters, *J. Pharm. Sci.*, 68, 1534, 1979.

47. **Yolles, S., Ward, L., Boetne, F., Leafe, T., Sartori, M., and Torkelson, M.**, Controlled release of biologically active agents, in *Controlled Release Polymeric Formulations, ACS Symp. Ser. No. 33*, Paul, D. R., and Harris, F. W., Eds., American Chemical Society, Washington, D. C., 1976, 123.

48. **Mason, N., Thies, C., and Cicero, T. J.**, *In vivo* and *in vitro* evaluation of a microencapsulated narcotic antagonist, *J. Pharm. Sci.*, 65, 847, 1976.

49. **Lewis, D. H., Dappert, T. O., Meyers, W. E., Pritchett, G., and Suling, W. J.**, Sustained Release of Antibiotics from Biodegradable Microcapsules, 7th Int. Symp. Controlled Release of Bioactive Materials, Ft. Lauderdale, Fla., July 28 to 30, 1980, 129.

50. **Yolles, S., Leafe, T., Ward, L., and Boettner, R.**, Controlled release of biologically active drugs, *Bull. Parenter. Drug Assoc.*, 39, 306, 1976.

51. **Wise, D. L., McCormick, G. J., Willet, G. P., Anderson, L. C., and Howes, J. F.**, Sustained release of sulphadiazine, *J. Pharm. Pharmacol.*, 30, 686, 1978.

52. **Yolles, S., Eldridge, J. E., and Woodland, J. H. R.**, Sustained delivery of drugs from polymer/drug mixtures, *Polym. News*, 1, 9, 1971.

53. **Pitt, C. G., Jeffcoat, A. R., Zweidinger, R. A., and Schindler, A.**, Sustained drug delivery systems. I. The permeability of poly(ϵ-caprolactone), poly (DL-lactic acid), and their copolymers, *J. Biomed. Mater. Res.*, 13, 490, 1979.

54. **Wise, D. L., Rosenkrantz, H., Gregory, J. B., and Esber, H. J.**, Long-term controlled delivery of levonorgestrel in rats by means of small biodegradable cylinders, *J. Pharm. Pharmacol.*, 32, 399, 1980.

55. **Pitt, C. G., Marks, T. A., and Schindler, A.**, Biodegradable drug delivery systems based on aliphatic polyesters: application to contraceptives and narcotic antagonists, in *Controlled Release of Bioactive Materials*, Baker, R. W., Ed., Academic Press, New York, 1980, 19.

56. **Brode, G. L. and Koleske, J. V.**, Lactone polymerization and polymer properties, in *Polymerization of Heterocyclics*, Vogel, O. and Furukawa, J., Eds., Marcel Dekker, New York, 1973.

57. **Pitt, C. G., Hendren, R. W., Schindler, A., and Woodward, S. C.**, The enzymatic surface erosion of aliphatic polyesters, *J. Controlled Release*, 1, 3, 1984.

58. **Bucher, J. E. and Slade, W. C.**, The anhydrides of isophthalic and terephthalic acids, *J. Am. Chem. Soc.*, 31, 1319, 1909.

59. **Hill, J. W. and Carothers, W. H.**, Studies of polymerization and ring formation. XIV. A linear superpolyanhydride and a cyclic dimeric anhydride from sebacic acid, *Am. Chem. Soc.*, 54, 1569, 1932.

60. **Conix, A.**, Aromatic polyanhydrides, a new class of high melting fiber-forming polymers, *J. Polym. Sci.*, 29, 343, 1958.

61. **Conix, A.**, Poly[1,3-bis(p-carboxyphenoxy)propane anhydride], *Macromol. Synth.*, 2, 95, 1966.

62. **Rosen, H. B., Chang, J., Wnek, G. E., Linhardt, R. J., and Langer, R.**, Bioerodible polyanhydrides for controlled drug delivery, *Biomaterials*, 4, 131, 1983.

63. **Leong, K. W., Brott, B. C., and Langer, R.**, Bioerodible polyanhydrides as drug-carrier matrices. I. Characterization, degradation, and release characteristics, *J. Biomed. Mater. Res.*, 19, 941, 1985.

64. **Leong, K. W., D'Amore, P., Marletta, M., and Langer, R.**, Bioerodible polyanhydrides as drug-carrier matrices. II. Biocompatibility and chemical reactivity, *J. Biomed. Mater. Res.*, 20, 51, 1986.

65. **Langer, R., Brem, H., and Tapper, D.**, Biocompatibility of polymeric delivery systems for macromolecules, *J. Biomed. Mater. Res.*, 15, 267, 1981.

66. **Choi, N. S. and Heller, J.**, U.S. Patent 4,093,709, 1978.

67. **Heller, J., Penhale, D. W. H., and Helwing, R. F.**, Preparation of poly(ortho esters) by the reaction of diketene acetals and polyols, *J. Poly. Sci. Polym. Lett. Ed.*, 18, 619, 1980.

68. **Heller, J., Penhale, D. W. H., Helwing, R. F., Fritzinger, B. K., and Baker, R. W.**, Release of norethindrone from polyacetals and poly(ortho esters), *AIChE Symp. Ser.*, 206, 28, 1981.

69. **Capozza, R. C., Schmitt, E. E., and Sendelbeck, L. R.,** Development of Chronomers™ for narcotic antagonists, *Natl. Inst. Drug Abuse Res. Monogr. Ser.,* 4, 39, 1976.
70. **Benagiano, G., Schmitt, E., Wise, D., and Goodman, M.,** Sustained release hormonal preparations for the delivery of fertility-regulating agents, *J. Polym. Sci. Polym. Symp.,* 66, 129, 1979.
71. **Heller, J.,, Penhale, D. W. H., Helwing, R. F., and Fritzinger, B. K.,** Release of norethindrone from poly (ortho esters), *Polym. Eng. Sci.,* 21, 727, 1981.
72. **Alfrey, R. T., Gurnee, E. F., and Lloyd, W. G.,** Diffusion in glassy polymers, *J. Polym. Sci.,* C12, 249, 1966.
73. **Fedors, R. F.,** Osmotic effects in water absorption by polymers, *Polymer,* 21, 207, 1980.
74. **Heller, J., Penhale, D. W. H., Helwing, R. F., and Fritzinger, B. K.,** Controlled release of norethindrone from poly (ortho esters), in *Controlled Release Delivery Systems,* Roseman, T. J. and Mansdorf, S. Z., Eds., Marcel Dekker, New York, 1983, 91.
75. **Heller, J.,** Controlled drug release from poly(ortho esters), in *Proc. 11th Int. Symp. Controlled Release of Bioactive Materials,* Meyers, W. and Dunn, R., Eds., Controlled Release Society Abstracts, 1984, 128.
76. **Sparer, R. V., Shih, C., Ringeisen, C. D., and Himmelstein, K. J.,** Controlled release from erodible poly (ortho ester) drug delivery systems, *J. Controlled Release,* 1, 23, 1984.
77. **Sanders, L. M., Kent, J. S., McRae, G. I., Vickery, B. H., Tice, T. R., and Lewis, D. H.,** Controlled release of a luteinizing hormone-releasing hormone analogue from poly(DL-lactide-co-glycolide) microspheres, *J. Pharm. Sci.,* 73, 1294, 1984.
78. **Wood, D. A.,** Biodegradable drug delivery systems, *Int. J. Pharm.,* 7, 1, 1980.
79. **Smith, D. C.,** Medical and dental applications of cements, *J. Biomed. Mater. Res.,* 5, 189, 1971.
80. **Vezin, W. R. and Florence, A. T.,** *In vitro* heterogeneous degradation of poly(n-alkyl α-cyanoacrylates, *J. Biomed. Mater. Res.,* 14, 93, 1980.
81. **Wood, D. A., Whately, T. L., and Florence, A. T.,** Formation of poly(butyl 2-cyanoacylate microcapsules and the microencapsulation of aqueous solutions of [^{125}I]-labelled proteins, *Int. J. Pharm.,* 8, 35, 1981.
82. **Hopfenberg, H. B.,** Controlled release from erodible slabs, cylinders and spheres, in *Controlled Release Polymeric Formulations, ACS Symp. Ser. No. 33,* Paul, D. R. and Harris, F. W., Eds., American Chemical Society, Washington, D.C., 1976, 26.
83. **Cooney, D. O.,** Effect of geometry on the dissolution of pharmaceutical tablets and other solids: surface detachment kinetics controlling, *AIChE J.,* 18, 446, 1972.
84. **Baker, R. W. and Lonsdale, H. K.,** Erodible controlled release system, *Am. Chem. Soc. Div. Org. Coat. Plast. Chem. Prepr.,* 3, 229, 1976.
85. **Heller, J. and Baker, R. W.,** Theory and practice of controlled drug delivery from bioerodible polymers, in *Controlled Release of Bioactive Materials,* Baker, R. W., Ed., Academic Press., New York, 1980, 1.
86. **Higuchi, T.,** Rate of release of medicants from ointment bases containing drugs in suspension, *J. Pharm. Sci.* 50, 874, 1961.
87. **Lee, P. I.,** Diffusional release of a solute from a polymeric matrix-approximate analytical solution, *J. Membr. Sci.,* 7, 255, 1980.
88. **Thombre, A. G. and Himmelstein, K. J.,** Modelling of drug release kinetics from a laminated device having an erodible drug reservoir, *Biomaterials,* 5, 250, 1984.
89. **Rosen, H.** Master's thesis, Massachusetts Institute of Technology, Cambridge, 1981.
90. **Goodman, M. and Kirschenbaum, G. S.,** U.S. Patent 3,773,737, 1973.
91. **Bamford, C. H., Elliot, A., and Hanley, W. E.,** *Synthetic Polypeptides,* Academic Press, New York, 1956.
92. **Randall, A. A.,** British Patent 1,049,290, 1966.
93. **Sidman, K. R., Schwope, A. D., Steber, W. D., Rudolph, S. E., and Poulin, S. B.,** Biodegradable, implantable, sustained release systems based on glutamic acid copolymers, *J. Membr. Sci.,* 7, 277, 1980.
94. **Marans, N. S.,** U.S. Patent, 3,111,469, 1963.
95. **Baptist, J. N.,** U.S. Patents 3,036,959, 1962, 3,044,942, 1962, 3,225,766, 1965.
96. **Shalaby, S. W. and Jamiolkowski, D. D.,** U.S. Patent 4,130,639, 1978.
97. **Casey, D. J. and Epstein, M.,** U.S. Patent 4,048,256, 1977.
98. **Schmitt, E. E. and Polistina, R. A.,** U.S. Patent 3,809,683, 1974.
99. **Capozza, R.C.,** Polymeric orthoesters, German Patent 2,715,502, 1977.
100. **Averbach, B. L. and Clark, R. B.,** The Processing of Chitosan Membranes, MITSG 78-14, MIT Sea Grant Program, Massachusetts Institute of Technology, Cambridge, 1978, 1.
101. **Maser, B., Cefelin, P., Lipatova, T. E., Bakalo, L. A., and Lugovskya, G. G.,** Synthesis of polyurethanes and investigation of their hydrolytic stability, *J. Polym. Sci. Polym. Symp.,* 66, 259, 1979.
102. **Doddi, N., Versfelt, C. C., and Wasserman, D.,** Synthetic Absorbable Surgical Devices of Poly(dioxanone), U.S. Patent 4,052,988, 1977.

Chapter 6

FABRICATING SILICONE RUBBER

Virgil L. Metevia and Patrick A. Walters

TABLE OF CONTENTS

I. INTRODUCTION: TYPES OF SILICONE RUBBER

Fabrication technology of silicone rubber is unusually versatile. Medication-impregnated parts, rate-controlling membranes, and reservoirs can be produced in almost any shape or size, allowing easy adaptation to other variables such as drug-diffusion rate, under-skin or on-skin application, and amount of medicament per unit.

Finished parts that are almost identical in appearance and properties can be made from silicone rubber formulations that, unvulcanized, range from easily pourable liquids to materials so viscous that they appear to be solids. A variety of curing or vulcanizing technologies are available — some involving heat of 95 to 200°C (203 to 392°F) and others that cure at room temperature. Drugs that are sensitive to heat can often be formulated with any of several of these room-temperature-vulcanizing (RTV) systems. Many types of silicone rubbers are intercompatible, allowing the creation of composite products to attain special goals. To achieve high, uniform release rates, for example, a coating of silicone rubber containing a maximum amount of drug can be applied to an inner core of a tougher, stronger grade of silicone rubber that provides optimum physical support. For high production rates, parts can be fabricated by high-speed injection molding or extrusion processes or die-cut from calendered sheets. Composite parts can be comolded, coextruded, or laminated. If lesser precision in dimensions is acceptable, composite parts may be produced by dip-coating.

In many cases, a stock medical-grade silicone rubber part such as a section of tubing or rod can be used in the fabrication of drug release packages. The drug itself or a silicone liquid or gel containing the drug can be injected into stock tubing, disks, or wafers of appropriate size and precise dimensions and then sealed in place with medical-grade silicone adhesive. Other methods are to fill tubing with liquid rubber containing the medicament and to cure it in place to extrude, mold, or coat a silicone rubber containing the drug on a section of silicone rubber tubing or rod.

A full understanding of the great variety of possible solutions must begin with a knowledge of the types of silicone rubber available as possible starting materials. This information can be integrated with other variables — including the heat and chemical stability of the drug, the desired mode of administration, diffusion rates, and many other variables — to select an appropriate fabrication technology. Then, thorough testing must be done to establish the safety and efficacy of the fabricating system. In particular, testing of drug-containing formulations must be done to establish that, under possible production conditions (including storage materials in process), (1) the drug does not adversely affect the curing reaction or properties of the silicone rubber formulation and (2) the curing agents, the physical action of the fabricating machinery, and/or the curing temperatures do not affect the composition of the medicament.

From the standpoint of fabrication, silicone rubbers can be classified by two characteristics: the consistency of the rubber in the unvulcanized state and the curing technology used to vulcanize it (see Table 1). Silicone rubber parts can be made with any of the following types of rubbers, to match most needs for medical-grade elastomers.

A. High-Consistency Silicone Rubber

High-consistency silicone rubbers, the first to be developed, are still the most important commercial silicone rubber materials for the fabrication of parts. In general, they are handled with the same type of equipment used in processing and fabricating natural rubber and many synthetic elastomers. The high-consistency silicone rubbers are mixed or blended in high-intensity mixers or on two-roll rubber mills; then they are resoftened on the two-roll mills. After that they are taken off the mills in sheets or "sheeted off". Finally, they are either die-cut into molding preforms for compression or transfer molding or they are cut into strips of raw elastomer for feeding into extruders or high-pressure injection-molding machines.[1]

Table 1
SILICONE RUBBERS USED IN DRUG RELEASE PRODUCTS

Type of product	Viscosity, uncured	Vulcanizing temperatures	Fabricating equipment
High-consistency			
Platinum-cured	Soft solid	95—200°C (203—392°F)	Two-roll mills and other
Peroxide-cured	Soft solid	116—170°C (241—338°F)	equipment used to fabricate ordinary rubber parts, plus oven for postcure
Low-consistency, two-part			
Tin curing agent	Easily pourable	Room temp	Hand mixable or can be handled in automated meter-mixers
Platinum curing agent	Easily pourable	Room temp, or can be accelerated by heat	
Platinum curing agent, gel	Easily pourable	Room temp to 150°C (302°F)	
Platinum curing agent	Semifluid	200°C (392°F) (lower temps possible)	Injection molders, extruders, HAVs
Low-consistency, one-part	Semifluid	Room temp	Hand applied from tubes or caulking guns or can be handled in automated equipment
Fluids and emulsions (not rubbers)	20—12,500 St	Not curable	Hand poured, gravity feeds, or pumps

High-consistency silicone rubbers use either of two types of vulcanizing agents: platinum compounds or organic peroxides. Silicone rubbers cured with platinum compounds require little or no postcure, since fewer volatiles are produced during vulcanization. Curing is accelerated by heat. The usual curing procedures call for heating for a period of time from about 2 min at 160°C (320°F) to approximately 8 min at 95°C (203°F). This addition-type curing reaction has many advantages, but it has the disadvantage that it may be inhibited by traces of amines, sulfur, organotin compounds, nitrogen oxide, and/or carbon monoxide.

In contrast, when used in medical applications, silicone rubbers vulcanized with organic peroxides usually require a postcure, which is carried out in an air-circulating oven. The reason is inherent in the mechanism of the cure. When the silicone rubber is heated to a specific temperature, the peroxides decompose to form reactive molecular fragments called free radicals. These free radicals activate reactive sites on the silicone polymer chains, causing them to cross-link with other silicone polymer chains to form elastomers. The temperature required for vulcanization with organic peroxides ranges from 115 to 170°C (239 to 338°F).

With either platinum compounds or organic peroxide vulcanizing agents, the heat necessary for vulcanization is often applied to the raw silicone rubber from hot molds or platens or — in the case of extrusions — in a heated tunnel called a hot-air-vulcanizer (HAV) unit. Some types of fabrications are vulcanized in an autoclave or a continuous-steam-vulcanizer (CV) unit, and still other methods are used in special cases.

Because many drugs would be affected by the elevated temperatures needed to cure these high-consistency rubbers, or by contact with organic peroxide curing agents, these materials are employed mainly to form tubing, rod, or membrane used to enclose or support the formulation containing the medicament.

B. Two-Part Low-Consistency Silicone Rubber
Two-part liquid silicone rubber formulations are very useful in fabricating controlled

release drug delivery systems. These materials are cured at room temperature or with moderate heating after a catalyst is added. The low curing temperatures used allow the rubber to be blended with heat-sensitive drugs. The pourable or semifluid consistency of these materials makes mixing easy. The semifluid formulations can be injection-molded, extruded, or formed into sheet stock by coating backing sheets with the mixture.

Platinum compounds and tin compounds are the most commonly used curing agents for the two-part low-consistency silicone rubbers. Cure rates can be varied from a few seconds to several hours by the choice of curing agent and the curing temperature. Many of these materials cure efficiently and quickly at room temperature. Since the performance of each of these curing agents depends on the exact formulation that is being used, it is advisable to consult a supplier of medical-grade silicone rubber before using any curing agent other than the one recommended for the formulation to be used.

Silicone gels, two-part materials that resemble a tough jelly in consistency, are also used as the matrix or carrier in drug release systems. They are enclosed in a silicone rubber tube or wafer or in some other enclosure with a silicone rubber rate-controlling membrane. The gels contain no fillers, and their cohesive yet conforming nature closely approximates the consistency of normal fatty tissue found in the body.

Noncuring silicone fluids in viscosities from 20 to 12,500 cSt (waterlike to heavy molasses in consistency) are also used as drug carriers or matrices. They are enclosed in the same manner as gels, but may offer higher release rates.

C. One-Part Low-Consistency Silicone Rubber

One-part silicone rubbers are supplied in the form of liquids or pastes and packaged in sealed tubes or cartridges. These materials cure at room temperature when they are exposed to air by reacting with moisture vapor. No heat is given off and ordinary levels of humidity are sufficient to accomplish this reaction.

The one-part silicone rubbers used in medical products are termed "acetoxy" materials. This terminology refers to the presence of occasional acetoxy groups on the ends of the siloxane polymer chains. Since the Si-O-C chemical linkaage is hydrolyzable, exposure to water vapor causes the acetoxy group to split off, resulting in cross-links which form the silicone rubber polymer matrix. The acetoxy by-product forms acetic acid, creating a mild vinegar odor until curing is complete. The reaction is nonreversible. The silicone rubbers formed are chemically similar to those formed by the use of the vulcanizing or curing agents described previously.

The chief advantages of one-part silicone rubbers are that they cure at room temperature and that no mixing or curing equipment is necessary for vulcanization. Potential disadvantages include a longer cure time, poor cure of thick sections, and the possibility that acetic acid formed during curing could react with a particular drug. These materials are used mainly as adhesives for silicone rubber and other materials.

II. FABRICATING HIGH-CONSISTENCY SILICONE RUBBER

A number of medical-grade silicone rubbers are available in unvulcanized form, essentially ready for molding, extruding, or handling by other manufacturing processes. For most applications, the maker of drug-diffusion materials can simply choose an appropriate formulation. In a few cases where particular properties are required, such as a higher degree of hardness, it is possible to modify the preformulated rubbers by the addition of fillers. Two courses of action are open to the manufacturer: to seek the necessary compounding advice from a silicone manufacturer experienced in medical-grade rubbers or to buy special grades of silicone rubber from a firm that specializes in compounding such elastomers.

A. Preparation

Silicone rubbers with platinum-based curing agents are supplied as two components, which must be uniformly cross-blended on the mill. One component, designated in the manufacturer's literature, is softened on the mill first and then removed. Then the second component is softened, after which the two components are thoroughly cross-blended, using procedures described in Section II.A.3. During this procedure, the temperature of the silicone rubber mut be kept as low as possible to prolong table life. A good rule is to blend only enough material for the next 3 to 4 hr of operation. However, if necessary, blended material can be stored in a freezer (below 0°C or 32°F) for as long as 7 days if it is carefully wrapped. When this material is removed from cold storage, it should be warmed to room temperature before the wrapping is removed to avoid condensation of water on the elastomer. Water can cause voids in molded or extruded parts.

The preparation of silicone rubbers containing peroxide vulcanizing agents is only slightly different from that used for the two-part platinum-cured high-consistency compounds. Freshly worked silicone rubbers that contain vulcanizing agents can be used directly in molding, extruding, calendering, and other processes. However, many raw silicone rubbers tend to stiffen and harden in storage, a phenomenon known as "crepe hardening". Crepe-hardened material must be resoftened before it is processed in fabrication machinery. If the raw silicone rubber being used does not contain vulcanizing agent, the organic peroxide can be added during this resoftening or masticating step. In medical manufacturing, these operations are usually carried out on a two-roll rubber mill.

1. Recommended Milling Equipment[2]

Silicone rubber is milled on a standard two-roll differential-speed rubber mill (See Figure 1). The mill is placed with the faster roll accessible to the operator. It should have:

1.   Water-cooled rolls (The rubber is so stiff that it may be heated by the physical working that it receives on the mill. The raw rubber should be kept at temperatures below 55°C (131°F) during this operation, since higher temperatures may cause premature vulcanization or partial volatilization (and loss) of the vulcanizing agent.)
2.   An adjustment mechanism on the fast roll which controls the bite (distance between the rolls)
3.   A peripheral speed of 75 to 150 ft/min on the fast roll
4.   A ratio of fast-roll speed to slow-roll speed in the range of 1.2:1 to 1.4:1
5.   A nylon scraper blade on the fast roll and nylon edge plows on both rolls (Nylon is preferred because it easily takes the shape of the roll, and it does not create metal wear particles that would contaminate the rubber being milled.)
6.   A pan under the rolls that can be easily removed, to catch material that drops from the rolls during milling (This pan must be kept free of dirt, grease, and oil.)
7.   Sealed or shielded bearings so arranged that no grease or oil can escape to contaminate the rubber

2. Milling the Rubber

To soften the raw rubber on the mill, the mill must be adjusted so that its bite or clearance is about 1/8 in. Then it must be loaded with an appropriate amount of rubber. A mill with 8-in.-diameter, 12-in.-long rolls will handle 1 to 2 lb of silicone rubber; a 12-in. by 24-in. mill, 12 to 14 lb of silicone rubber; a 20-in. by 60-in. mill about 60 lb; and a 35-in. by 84-in. mill about 120 lb.

If the rubber has become very stiff and is not easily pulled into the bite when the mill is operated, the bite may be opened slightly or the material cut into thin pieces. Pieces that drop through the bite into the collection pan should be put back in the mill immediately,

FIGURE 1.   Two-roll rubber mill typical of those used for silicone rubber.

provided they have not become contaminated with dirt or grease. Hard material should not be added to soft or partly softened material. It will not mill in smoothly and may produce irregularities in the finished product. Very hard rubber may band more easily on the rolls if, after the rubber has gone through the rolls a few times, the bite is set to 1/16 in. for a short time and then opened to about 3/16 in. Silicone rubber usually bands on the slow roll at first. As the rubber is softened, it transfers to the fast roll. When milling is nearly complete, all the rubber will be on the fast roll.

3. Cross-Blending

Cross-blending is used to blend the two parts of the platinum-cured formulations and ensure complete mixing when adding the vulcanizing agent to peroxide-cured formulations. With both platinum-curable and peroxide-curable silicone rubbers, the technique is used to make sure that a compound is uniformly softened. It is also used to blend two different silicone rubbers to obtain formulations with properties intermediate to those of the starting materials.

To cross-blend, pieces of rubber from the fast roll must be stripped with the scraper blade, turned 90°C, and fed back into the mill. They must be started at one end of the bite and guided along the rolls. If the mill is not operated properly, air can be trapped in the rubber during milling, and a popping noise will be heard as the mill is operated. Entrapped air will cause problems in fabrication. To avoid such problems, the mill should be set to a 1/16- or 1/8-in. bite when the rubber transfers to the fast roll and just enough rubber should be milled to have only a slight bank visible above the bite. As rubber is taken from the fast roll during cross-blending, a small bank will form, but it will disappear as the rubber works out to the full length of the rolls.

Table 2
ORGANIC PEROXIDE VULCANIZING AGENTS USED IN MEDICAL-GRADE SILICONE RUBBERS

Vulcanizing agent	Typical molding conditions for 1/8-in. thick moldings	Remarks
bis-2,4-Dichlorobenzoyl peroxide	5 min at 115°C (230°F)	Good for continuous hot-air vulcanization, scorch may be a problem in thin section
Benzoyl peroxide	5 min at 127°C (260°F)	Good for thin moldings and for dispersion coatings; not recommended for HAV
2,5-Dimethyl-2,5-di (*t*-butyl-peroxy) hexane	10 min at 171°C (340°F)	Gives good all-around properties and low compression set; good for thick-section cures
Dicumyl peroxide	10 min at 160°C (320°F)	Good for thick-section cures; provides low compression set; used if lower cure temperatures than those required by preceding agent are needed

On a 12-in. roll, most rubber compounds will be completely softened in 5 to 8 min after the rubber transfers to the fast roll. With larger rolls, the time may be longer. Incompletely milled silicone rubbers will flow poorly in molds or extruder dies, resulting in defective parts. Extruded parts may have a rough or "orange peel" surface. Milling too long may make the rubber sticky and hard to handle and may affect the fabrication of parts made from platinum-curable compounds.

4. Pigmenting

Silicone rubbers can be formulated in various colors by adding inorganic pigments. However, the formulator must establish that the pigment to be used is safe for the intended purpose. Some pigments are toxic and should not be used for medical-grade rubbers.

B. Vulcanizing

1. Vulcanizing Agents

In drug diffusion technology, the most widely used vulcanizing agents are the platinum compounds. These curing agents vulcanize only silicone rubbers which contain vinyl groups. Because platinum compounds cure by an addition-type reaction, there are no volatile by-products. Curing takes place at fairly low temperatures: about 7.5 min at 93°C (200°F), for example, or about 1.5 min at 160°C (320°F). Vulcanization time will vary depending on the exact formulation, but it is generally controlled by adjusting the temperature.[3]

Platinum-cured silicone rubbers have few disadvantages. The main limitation is that this curing mechanism may be inhibited by traces of amines, sulfur, nitrogen oxide, organotin compounds, carbon monoxide, and residues from curing agents from RTV silicone elastomers and peroxide-cured elastomers. Equipment and compounds must be free of these substances.

Organic peroxide vulcanizing agents used in silicone rubbers in drug diffusion technology are shown in Table 2. Of these, the most commonly used vulcanizing agent is *bis*-2,4-dichlorobenzoyl peroxide. It vulcanizes all types of high-consistency silicone rubber. It is commonly used in compounds intended for extrusion, since it vulcanizes effectively in continuous HAV equipment, with little tendency to produce porosity in parts that are vulcanized without external pressure. However, in thin molded parts, scorch (partial vulcanization before flow is complete) may be a problem. Vulcanization with *bis*-2,4-dichlorobenzoyl peroxide takes place at approximately 115°C (239°F).

Another vulcanizing agent, benzoyl peroxide, also vulcanizes all types of silicone rubbers. It is activated at approximately 125°C (257°F). It has little tendency to produce scorch in thin moldings, but it is not recommended for continuous HAV, since it may produce porosity in unconfined silicone rubber. This vulcanizing agent is often used in the silicone rubber dispersions used in coating cloth, since it vulcanizes the rubber coating efficiently in the curing towers often used with this fabricating method. However, the amount of peroxide needed for dispersions ranges from four to ten times the amount needed for molding.

A third organic peroxide often used as a vulcanizing agent for high-consistency medical-grade elastomers is 2,5-dimethyl-2,5-di(*t*-butylperoxy)hexane, which is activated at about 170°C (338°F). Compounds containing this agent may be prepared up to 8 weeks in advance of use, with no need to resoften the compounds. This vulcanizing agent gives rubbers excellent physical properties and is especially suitable for parts with thick cross sections. However, it is normally used only with silicone polymers which contain vinyl groups.

Dicumyl peroxide is useful in the special case where vulcanization equipment is limited to a maximum temperature lower than that required for 2,5-dimethyl-2,5-di(*t*-butylpe-roxy)hexane, but where the advantages of good thick-section cure and low-compression set are desired. This organic peroxide vulcanizes silicone rubbers which contain vinyl groups at 160°C (320°F). The main disadvantage of this peroxide is that it requires a longer postcure than other peroxides to remove volatiles from the silicone rubber. Several other organic peroxides may be used as vulcanizing agents, but the first three listed above are the most widely employed in medical-grade elastomers.

2. Amount of Vulcanizing Agent

The optimum amount of vulcanizing agent for silicone rubber in a production operation depends on two variables: the formulation and the fabricating method. Some formulations contain vulcanizing agents. In the case of platinum-catalyzed compounds, two components are supplied. Other formulations are designed to be vulcanized with a peroxide vulcanizing agent, but are supplied without it. With these formulations, the user must choose a specific peroxide vulcanizing agent and add it as part of the manufacturing operation. Data sheets on the formulation will generally supply suggestions as to the type and amount of vulcanizing agent needed to cover most fabricating operations.

3. Causes and Cures of Vulcanizing Problems

Scorch can occur with either platinum- or peroxide-vulcanized compounds. With platinum-cured compounds, it may be produced by overheating the compound on the mill. With either type of vulcanizing agent, it may be caused (in molding) by hot spots in the mold, molding at too high a temperature, or slow flow rates in the mold. If finished products are weak, soft, cheesy, or gummy, the formulation may have had insufficient vulcanizing agent. With platinum-based vulcanizing agents, such a condition might be caused by poor blending of the two components of the formulation or by the presence of a material that inhibits the curing agent. Inhibition may be produced by trace quantities of a number of materials, as indicated in Section II.B.1. In rubber formulations containing organic peroxide vulcanizing agents, the peroxide may have been lost by volatilization or it may have been consumed by preferential side reactions without accomplishing the cross-linking of the rubber. Volatilization of peroxides can result from:

1.   Molding (or hot-air vulcanizing, etc.) at too low a temperature — This is the most common cause for poor vulcanization. The organic peroxide is "boiled off" before it reaches its activation temperature.
2.   Allowing preforms or other thin sections of unvulcanized silicone rubber to stand in warm areas for long periods — Some of the peroxide evaporates or sublimates from the rubber, leaving too little to accomplish full vulcanization.
3.   Delay in starting the press after loading a heated mold — This may result in scorch (premature vulcanization) or loss of vulcanizing agent from evaporation.

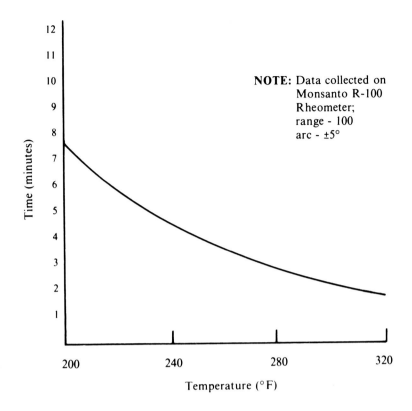

FIGURE 2.   Cure rate of a platinum-cured medical-grade silicone rubber.

"Preferential reactivity" is the term applied to situations in which the silicone rubber contains some substance that reacts more easily with the peroxide than does the silicone rubber polymer. Preferential reactivity can be caused by entrapped air, various solvents, or other organic ingredients. Corrective measures for these problems can include changes in the fabrication method or work procedure or changes in ingredients. Where these solutions are not practical, it may be possible to correct the problem by increasing the amount of organic peroxide vulcanizing agent. Then, even though some peroxide may be lost, enough will remain to do the job. In other cases, a change in vulcanizing agent may eliminate the problem.

4. Effect of Temperature and Vulcanizing Agent on Viscosity

When a silicone rubber containing a vulcanizing agent is heated, its viscosity decreases with increasing temperature until vulcanizaion begins. This effect is shown in Figures 2 and 3. Note that the shapes of the curves are altered by the choice of vulcanizing agent. Platinum-cured compounds cure over a fairly wide range of temperatures. The cure with organic peroxides occurs rapidly at a certain activation temperature. Those peroxides that require higher temperatures for activation allow the rubber to become softer before vulcanization is initiated. A rubber containing 2,5-dimethyl-2,5-di(*t*-butylperoxy)hexane, for example, will reach a lower viscosity (and thus flow better in the mold) than the same rubber using *bis*-2,4-dichlorobenzoyl peroxide, which is activated at a lower temperature.

5. Vulcanization of Thick Sections

Vulcanization of thick sections of silicone rubber presents two problems: more time is required for heat to penetrate to the center of the rubber and effect vulcanization and more

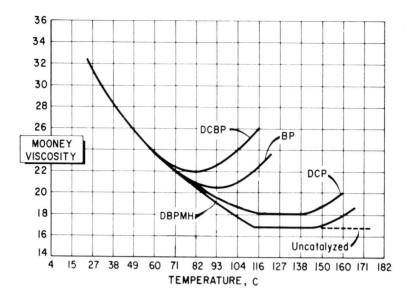

FIGURE 3.   Cure rates of silicone rubber with various organic peroxide vulcanizing agents.

time is required during postcuring for volatiles to diffuse through the rubber and escape from the outer surfaces. Postcuring is discussed later. For silicone rubber of 1/8-in. thickness, the postcure suggested in the product data sheet is generally sufficient. Thicker sections should be given longer cures.

C. Molding

Three principal methods are used to mold silicone rubber: compression molding, transfer molding, and injection molding. With each of these methods, the polymer device is formed and then vulcanized by heat and pressure in the mold.

Compression molding is the most widely used method. It is the oldest method, uses the simplest equipment, and can be used to produce many types of parts. Transfer molding is often used for short-run production of complex parts. Injection molding is used for high-volume production. All three processes make use of the same type of equipment which is used to handle organic rubbers. However, some modifications may be necessary to accommodate the special characteristics of silicone rubber.

D. Extruding

Extrusion of high-consistency silicone rubber uses the same principles as the extrusion of other types of rubber. The unvulcanized rubber is forced through a die that forms it into a continuous ribbon, rod, tube, or other predetermined cross section. It is then vulcanized and often postcured.

Silicone rubbers flow easily through the extruder and expand as they leave the die (as do organic rubbers), but they show less strength in the unvulcanized state than organic rubbers. For this reason, they must be vulcanized immediately after extrusion. Unlike some organic rubbers, silicone rubbers are extruded at room temperature. They need not be heated to attain adequate flow in the die. In fact, heating above 55°C (131°F) may cause a loss of vulcanizing agent or even scorch (premature vulcanization).

1. Preparing Rubber for Extruding

As in molding, silicone rubber should be softened on a rubber mill before extrusion. As with all other processing methods, cleanliness is very important. To keep the milled sheets clean and to prevent them from sticking together, they can be covered and separated with polyethylene sheets. To make it easy to feed the extruder uniformly, the unvulcanized rubber should be cut into strips about 1/2 inch. thick and 1 to 2 in. wide. These strips can be hand-fed or they can be fed from a "hat" as described in Section II.D.5.

2. Extrusion Equipment

Extruders for silicone rubber should have the following features:

1.   A screw designed for silicone rubber
2.   An extended barrel, matched to the length of the screw
3.   A breaker plate (perforated disk) to remove entrapped air (this plate should have a recess to hold screens)
4.   An 80- to 200-mesh screen (If a 200-mesh screen is used, it should be backed up with an 80-mesh screen to prevent the finer screen from rupturing. The screens can be welded together to better resist the high pressures which will develop.)
5.   A spider flange to hold mandrels used in making tubing, to mount the die holder, and to permit radial adjustment for centering
6.   A universal die holder
7.   A coin or plate die

3. Screw Design

For most silicone rubbers, a single-flight screw with diminishing pitch is used. Good results have been obtained with a screw having a length-to-diameter (L/D) ratio of 12:1, although screws with a 10:1 L/D ratio are also used. The latter screws tend to run cooler. Flights should be quite deep.

In some cases, rubber parts fabricators have trouble with dark streaks in silicone rubber extrusions, caused by metal wear particles. To prevent this, nylon plugs can be set radially in the forward end of the flight, to act as bearings that keep the metal flight from rubbing against the barrel. To make these nylon plugs effective, the outside diameter of the flight must be tapered.

4. Mandrels

In extruding tubing, the inside diameter of the tubing is formed by a mandrel, held in place by a spider flange. Makers of thin wall tubing drill the flange and mandrel to admit air under light pressure (1 to 5 psig) to keep the tubing round during vulcanization.

Mandrels are made with interchangeable tips, one for the inside diameter of each size of tubing. In extruding rubber with poor flow characteristics and in making close-tolerance tubing of greater than a 5/16-in. inside diameter, a ball tip helps in holding the diameter within tolerances.

5. Roller Feed

To save labor and ensure steady feeding of the extruder, long strips of silicone rubber are often fed from a hat-shaped fixture which rotates on a vertical, ball-bearing axle mounted close to the barrel. The strips pass over a roller mounted parallel with the screw. The clearance between roller and screw should be about 0.2 in. The roller should be fitted with an adjustable scraper and should be easily removed for cleaning.

6. Die Design

Coin or plate dies work well for extruding high-consistency silicone rubber. Good practice in die design involves consideration of several principles:

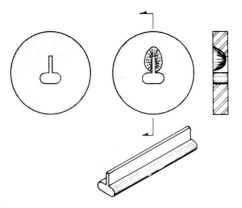

FIGURE 4.   Die for bulb-shaped extrusion has a short land machined out at the edges of the thin part of the die opening. This provides the same flow rate through the thin section as through the large part of the opening.

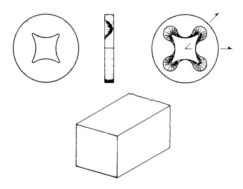

FIGURE 5.   Die for square extrusion. The sides of the die are convex so that the extrusion will have straight sides because the rubber "swells" as it leaves the die. A slight radius on the corners helps to make smooth corners on the extrusion

1.   Flow of silicone rubber should be free and uniform as it approaches the die. Protrusions or irregular surfaces may cause a "dead spot" which will trap and hold silicone rubber. If this rubber begins to vulcanize in the extruder, it will produce rough extrusions or other problems.
2.   Silicone rubber usually swells as it leaves the die. Die openings ordinarily must be somewhat smaller than the size desired in the extrusion, although shrinkage during vulcanization and postcuring partly compensates for die swell. The cross section of the extrusion can also be reduced somewhat by adjusting the speed of the extrusion as it passes to the HAV so that it stretches the unvulcanized extrusion slightly.
3.   Flow of silicone rubber should be uniform in all parts of the die. Sharp edges in the die tend to create excessive drag on the rubber, resulting in rough extrusions. This problem can be overcome by rounding all edges slightly. Nonuniform flow can also result when some parts of the die opening are smaller than other parts in cross-sectional area. To avoid this problem, improve flow at the constriction by machining away part of the metal (see Figures 4 and 5). Conversely, the flow through large openings can be slowed by installing a dam on the screw side of the die (see Figures 6 and 7).
4.   Dies should be designed to allow easy adjustments.

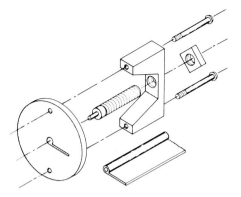

FIGURE 6. This die for a p-shaped extrusion features a means of adjusting flow. The hole in the "p" is formed by a pin mounted on a bridge. The flow in the thick and thin sections of the die is balanced by the adjustable shoulder dam behind the pin.

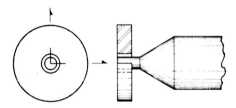

FIGURE 7. Die and mandrel for extruding silicone rubber tubing. Mandrel tip is replaceable for changing the inside diameter of the tubing.

E. Vulcanizing Methods and Equipment

1. Hot-Air Vulcanizing

Most silicone rubber extrusions are vulcanized by hot-air vulcanizing. In this continuous process, the extrusion is passed through a horizontal or vertical[4] chamber heated to 315 to 425°C (600 to 800°F). Extrusions with small cross sections require only a few seconds to vulcanize at these temperatures, and those of larger cross sections require proportionately longer times. Overcuring can produce surface hardening and even brittleness or crazing in the worst cases.

In horizontal HAV units, the extrusion is carried on a wire-mesh belt from the extruder through the vulcanizing chamber. In vertical units, the extrusion is pulled through the chamber by a motor-driven drum at the top. The extrusion is sufficiently vulcanized in the bottom portion of the chamber to prevent excessive stretching, and as it moves up through the chamber, it continues to vulcanize, gaining strength enough to hold the weight of the material below with little change in dimensions. With both horizontal and vertical types, the length of the unsupported loop of extrusion between the extruder and the vulcanizing chamber should be held constant to maintain a uniform cross section.

Vulcanizing chambers of either vertical or horizontal units can be heated by strip heaters, Calrod heaters, infrared units, or any other clean heat source.

All vulcanizing chambers must have an air-exhaust system to assure proper removal of volatiles. If platinum-based vulcanizing agents are used, these volatiles will be almost entirely low molecular weight silicones. If peroxide vulcanizing agents are employed, the composition of emissions will consist of low molecular weight silicones and — of greater concern —

products of the breakdown of the particular peroxide used. It is good practice to determine the exact composition of emissions. For more information on these breakdown products, consult the manufacturer of the vulcanizing agent. Check local, state, and federal regulations for details of allowable equipment and permissible volatiles in the work area and emissions to the environment.

Horizontal HAV units are best suited to extrusions that have one or more flat sides. Vertical units are best for making thin-walled tubing, since they leave no belt marks and have no tendency to flatten the extrusion. In addition, they provide more uniform heating around the extrusion, assuring uniform vulcanization. At the same temperature and air circulation rate, a vertical unit vulcanizes about 25% faster than a horizontal unit for a given length of vulcanizing chamber.

Vulcanizing speed can be increased by as much as 100% by providing hot-air circulation in the vulcanizing chamber. One method involves supplying air at about 10 psig to stainless steel tubes running along the walls of the chamber. The tubes lead this air to a center tube, where it is blown through holes in the tube walls into the chamber. This arrangement greatly improves heat transfer as compared to static chamber atmospheres.

2. Vulcanizing in Autoclaves

Short lengths of extruded silicone rubber can be vulcanized in autoclaves. The unvulcanized extrusion is loosely coiled in a pan or tray and subjected to sufficient steam pressure to provide the required temperature/time conditions for vulcanization.

F. Calendering

Calendering is used to produce long, thin sheets of silicone rubber of uniform thickness. It can also be used to produce fabric-backed sheets of silicone rubber. Either three-roll or four-roll calenders can be used, although four-roll units are more effective in keeping air bubbles out of the sheet. The calender should have a variable-speed main drive to give a center-roll speed range of 2 to 10 surface ft/min. This roll should be geared so that it runs about 1.2 to 1.4 times the speed of the top roll. The bottom roll runs at the same speed as the center roll. Nylon plows are used on all but the bottom roll to keep the rubber from creeping over the ends of the rolls.

Silicone rubber is usually calendered at room temperature. However, a means of heating the center roll should be provided, since some rubbers process better when the center roll is slightly heated. To avoid risk of scorching the rubber, the roll temperature should not go to above 55°C (131°F). Most silicone rubber must be softened on a mill before calendering. If the rubber tends to be soft and sticky, it should stand for 24 hr after milling before it is placed in the calender.

An unsupported silicone rubber sheet is generally calendered on a plastic sheet (such as Kodacel® from Eastman Kodak). Then plastic and rubber are wound on a hollow steel or aluminum core to a depth of about 5 in. This assembly is then wrapped tightly with wet cotton or nylon tape. The wrap is started at the middle of the roll and half-lapped to one end and then wrapped to the other end and back to the middle. The roll is then inserted in an autoclave at the temperature required for the particular vulcanizing agent being used. After vulcanizing, the rubber sheet is stripped from the Kodacel® sheet while it is still warm.

G. Postcuring

Postcuring of silicone rubber parts is done for two reasons: to remove unwanted volatiles and to stabilize the properties of the elastomer. Silicone rubber parts vulcanized with platinum-based curing agents often need no postcure. When a postcure is given, it is done to stabilize properties. Postcuring "ties up" any residual reactive sites in the molecule, producing additional cross-links between molecules. Postcures of 2 to 4 hr at 175°C (347°F) are usually sufficient.

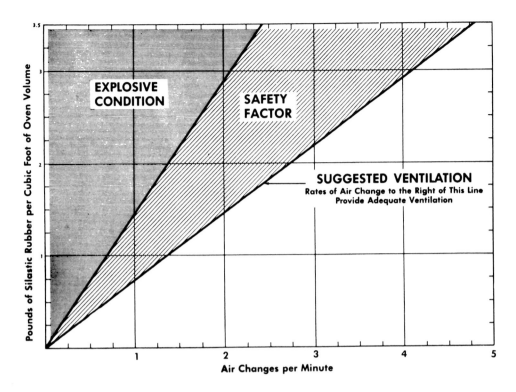

FIGURE 8. Graph shows explosive load limit for any oven used for postcuring silicone rubber.

Peroxide-vulcanized silicone rubbers used in medical applications must be postcured. Some of the volatiles formed by the decomposition of the organic peroxide vulcanizing agent are still present in the rubber after the initial vulcanization step. The postcure removes these unwanted breakdown products, making the elastomer more acceptable from a biological standpoint and, as with platinum-cured elastomers, stabilizing the properties of the rubber. The schedules for oven curing depend on the rate at which volatiles can escape from the silicone rubber. This, in turn, depends on the thickness of the part being cured and the amount of exposed surface. For instance, a 1/2-in. slab bonded on one side to metal or another impermeable material will require about the same cure schedule as a 1-in. slab with both surfaces exposed. A recommendation on the postcure or oven cure needed for each peroxide-vulcanized silicone rubber formulation is ordinarily supplied by the manufacturer. A typical postcure for thin sections would be 2 hr at 150°C (302°F). Thicker sections might require a postcure schedule of 2 hr at 150°C, followed by up to 12 hr at 200°C (392°F).

Curing ovens must be equipped with an exhaust system (see Figure 8) to remove volatiles which might otherwise reach explosive levels. Ovens also used an air-circulating fan to assist in heat transfer and to keep conditions uniform throughout the oven. Temperature controls are necessary to maintain proper oven heat, and an automatic safety switch should be provided to shut off the oven if the exhaust fan fails. As with HAV units and curing towers, manufacturers should determine what volatiles are present in the workplace and what emissions are going into the atmosphere and make sure that the operation meets all local, state, and federal regulations.

To avoid distortion of parts, they should not be stacked or placed in contact with each other. In curing tubing with wall thicknesses greater than 1/16 in., it may be necessary to circulate air through the tubing to avoid reversion of the peroxide-vulcanized silicone rubber. This precaution is not necessary, however, with platinum-cured elastomers.

Table 3
CURE TIME[a] OF A
TYPICAL LIQUID
SILICONE RUBBER
USED FOR INJECTION
MOLDING

Temperature	Cure time
−5°C (23°F)	3 months
50°C (122°F)	2 hr
75°C (167°F)	15 min
110°C (230°F)	110 sec
150°C (302°F)	50 sec
177°C (351°F)	20 sec
200°C (392°F)	5 sec

[a]   Cure measured using Monsanto
      rheometer.

III. FABRICATING TWO-PART LOW-CONSISTENCY SILICONE RUBBER

Two-part low-consistency silicone rubbers are supplied as two-liquid components that cure to form typical silicone rubbers when the two components are mixed. Neither component contains solvents. The liquid state of the components makes it easy to blend medicaments into the formulation. Automated mixing equipment is available for either continuous or batch-type blending of the two basic components, and drugs can be mixed into the formulation at the same time. Exact cure rates depend on the catalyst used and the curing temperature.

A. Vulcanizing

RTV silicone rubbers catalyzed with tin catalysts cure rapidly at room temperature. The cure reaction with tin catalysts is less sensitive to inhibition than for products which use platinum-based catalysts. These pourable RTV materials can be cast in simple forms or molds, making them particularly valuable for experimental work.

Two-part low-consistency silicone rubber formulations catalyzed with platinum-based curing agents are also available in pourable viscosities. They can be cured at room temperature or curing can be accelerated with heat. Like the tin-catalyzed RTV silicone rubbers, they are easily cast in simple forms or molds. They are particularly useful as drug matrices, since their room-temperature cure minimizes the chances of thermal breakdown of the drug. However, their cure is subject to inhibition from traces of amines, sulfur, organotin compounds, and certain other materials.

Semifluid formulations catalyzed with platinum-based curing agents are very well suited for high-volume production of high-quality parts by liquid-injection molding or extruding.[5-7] Injection molding is done in slightly modified machines developed for molding thermoplastics. Use of this production method with these low-consistency silicone rubbers eliminates the mill-softening and preform steps necessary with high-consistency silicone rubbers, yet their properties are often similar to those of general-purpose high-consistency silicone rubbers. Mold-cyle times can be as low as 10 sec or less (see Table 3).[8] With well-designed molds, the need for deflashing of parts is minimized or eliminated, and an operator may be necessary only for monitoring the automated mixing and fabricating operations, resupply of materials, startup, and shutdown.

From an economic standpoint, the production of silicone rubber devices from low-consistency silicone rubbers often offers significant advantages over operations involving high-consistency silicone rubbers. The capital costs of equipment are generally less when using low-consistency rubbers. Labor costs are much lower, and output rates are much higher.

Modifications of the thermoplastic injection-molding machines and methods are few, but significant. Because these silicones are lower in viscosity than thermoplastics, injection pressures are lower.[9] Mold tolerances must be closer than for thermoplastics to minimize flash on molded parts. For highest production rates, molds are heated to 150 to 200°C (302 to 392°F). Generally, molds should be designed with as little mass as possible so that they can be heated rapidly and efficiently. Placement of heating elements in the mold should be carefully considered, so that the mold surfaces are heated uniformly. Molds operated at 200°C (392°F) should be made of hardened steel for good tool life.

Another difference between silicone rubbers and thermoplastics is in the ejection of parts. Because silicone parts are more rubbery than thermoplastics, they cannot be ejected from the molds in exactly the same way. Yet cycle times are so rapid that part ejection must be quick and reliable. Modifications for part ejection include the provision of compressed air and mechanical assists if the molds are deep or complicated. Platinum-catalyzed silicone rubbers are also well suited for extrusion or coating operations, including the coextrusion technique.

B. Extruding
1. Extrusion of Liquid Silicone Rubbers

Extruding or two-part low-consistency silicone rubbers is a more easily automated procedure than the extrusion of high-consistency silicone rubbers. Silicone rubbers using platinum-based catalyst systems are used, and the two components are mixed in commercially available automatic metering/mixing equipment. Presently, low-consistency rubber extrusion is done over some type of support, such as fabric sleeving, a wire or thread, or a previous extrusion of silicone rubber. The support component is passed through the cross-head of the extruder, centered by guides in the cross-head, and the silicone rubber flows around it in the die. The composite then is passed immediately into a HAV. Curing is very rapid. Extrusion speeds of 200 ft/min are used in wire coating, with vulcanizing temperatures of about 315°C (600°F). Coating thicknesses of less than 0.1 mm to greater than 2.0 mm can be attained without difficulty.

Extrusion systems with these silicone rubbers require a continuous flow of the liquid rubber from the metering/mixing equipment. This flow is usually provided by the metering/mixing equipment itself, but it can also be done by using commercial gear pumps or pressure pots. Extrusion-head pressures are very low.

2. Extrusion over Prevulcanized Silicone Rubber Rod

By using a special extruder head, a precisely dimensioned coating of two-part low-consistency silicone rubber can be extruded over a silicone rubber rod that has been previously extruded and vulcanized. The technology used is an adaptation of techniques used in forming the insulating jacket on electrical wires. A guider tip (see Figure 9) keeps the silicone rubber rod (or other shape) centered in the die, and the liquid silicone rubber is pumped into a chamber that supplies material continuously to the die. The coated rod is then pulled through the vertical HAV tower and onto a take-up reel.

C. Coextrusion

Coextrusion of liquid silicone rubbers is also possible. By this means, a silicone rubber containing the drug can be continuously enclosed in a tubing of silicone rubber containing the medicament. The technique is an adaptation of the procedure described in the previous section. The difference is that two streams of premixed two-part liquid silicone rubbers are fed through two pressure chambers to two dies. One of the dies forms the primary or center

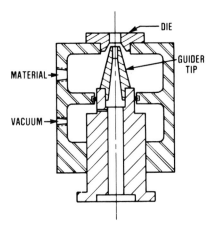

FIGURE 9.   Guider tip in extruder cross-head centers a thread, wire, or previous extrusion of silicone rubber that supports the coating of liquid silicone rubber until it enters the HAV.

extrusion over a supporting medium such as sleeving, wire, or thread; the other forms the outer cover.

The composite extrusion passes directly into a HAV from the extrusion head, and the two silicone rubber formulations are cured in the same pass. Because these silicone rubbers are very soft in the extruded form, the source of heat in the HAV must be as close to the extruder as is possible without heating the die enough to cause premature vulcanization.

IV. FABRICATING ONE-PART LOW-CONSISTENCY SILICONE RUBBER

Another group of silicone rubbers available for the fabrication of drug release materials is the one-part RTV silicone rubbers. These are soft, sealant-consistency silicone rubbers that vulcanize at room temperature on contact with water vapor in the air. The cure reaction involves moisture-sensitive chemical groups attached at wide intervals to the silicone molecules. When these groups contact water vapor, they split off from the silicone molecules, leaving reactive sites which are then occupied by cross-linkers that tie the silicone molecules together in a network. This network is a silicone rubber with much the same properties as silicone rubbers formed by other cross-linking methods.

The ease of cure of the one-part RTV silicone rubber makes it a logical candidate for the formulation of drug release products. The medicaments can be blended into the RTV rubber under airless conditions and the rubber can be used to make parts or coatings by simply applying it to the desired surface and exposing it to the air. The cure rate of the rubber can be increased somewhat by using a humidity chamber or by otherwise raising the water content of the air surrounding the rubber.

No heat is involved in the cure reaction, but acetic acid is formed as a transitory by-product of the reaction. This weak acid, which can be identified by its vinegar-like odor, quickly volatilizes into the surrounding air. Since the reaction depends on the diffusion of water vapor through the silicone rubber, the cure rate is dependent on the thickness of the section involved.

This type of silicone rubber, unlike most other types of silicone rubber, forms adhesive bonds to other materials that it contacts as it cures. For this reason, it has been used for many years as a medical-grade adhesive for silicone rubber parts, including some implantable prostheses. It is sometimes used as a closure for sections of silicone rubber containing fluids, gels, or other materials containing medicaments or to laminate sheets of silicone rubber in making patch-type drug-release products.

V. CONCLUSIONS

The fabrication technology available for use in making controlled release drug systems from silicone rubber is unusually versatile. A variety of curing systems are available to match silicone rubbers with medicaments, even with heat-sensitive or chemically sensitive medicaments. The design of release packages to fit differing drug delivery methods is also made easy by the variety of fabricating methods available. For experimental or prototype work, silicone rubber drug release systems can be fabricated without special equipment at the laboratory bench.

REFERENCES

1. **Lynch, W.,** *Handbook of Silicone Rubber Fabrication,* Van Nostrand Reinhold, New York, 1978.
2. Fabricating with Silastic® Silicone Rubber, Bull. No. 17-053b-79, Dow Corning Corporation, Midland, Mich., 1979.
3. Silastic® Medical Grade Elastomers, Bull. No. 51-582-82, Vol. 1, Dow Corning Corporation, Midland, Mich., 1982, 3.
4. Silicone Goods Improved by Vertical Curing, *Rubber Plast. News,* p. 6, January 10, 1977.
5. **Hays, W. R., Kehrer, G. P., and Monroe, C. M.,** Fabricating with liquid silicone rubber, *J. Elastomers Plast.,* 10 (2), 163, 1978.
6. **Hays, W. R. and Hilliard, J. R.,** A New Liquid System for Fabricating Silicone Rubber, Tech. Pap., Regional Technical Conference Society of Plastics Engineers, Milwaukee, 1980.
7. **Fraleigh, R. M. and Kehrer, G. P.,** Fabricating wire and cable with liquid silicone rubber, *Rubber World,* 181, (5), 33, 1980.
8. Liquid Polymer System, Bull. No. 17-307-80, Dow Corning Corporation, Midland, Mich., 1980, 5.
9. **Hays, W. R.,** Liquid silicone rubber processing by injection molding, extrusion, *Elastomerics,* 112, (2), 17, 1980.

Chapter 7

MOLDING OF CONTROLLED RELEASE DEVICES

W. Eugene Skiens

TABLE OF CONTENTS

I. INTRODUCTION

Over the past 14 years, research and development have been carried out to design, fabricate, test, and evaluate (both in vitro and in vivo) controlled drug release vaginal, cervical, and uterine contraceptive devices. These devices have been designed to release a variety of bioactive contraceptive agents which include steroids and spermicides. They have been developed for testing in a variety of animals as well as humans. A variety of molding techniques have been used in the development of these devices.

Since in many cases the effective levels of drugs to be delivered in the reproductive tract were unknown, fairly wide experimental release levels were developed for initial experimentations. A variety of polymeric materials have been used in these studies; however, in many instances, medical-grade silicone rubber has been selected as the material of choice for the drug delivery reservoir and releasing media. Another general type of polymer widely used has been olefinic polymers and copolymers.

The sponsors of this work have been primarily the World Health Organization, the National Institutes of Health, and the U.S. Agency for International Development. A number of research institutes, universities, and industrial organizations have acted as contractors in the design and fabrication of devices for these sponsoring institutions.

II. VAGINAL RINGS

Vaginal rings releasing both natural and synthetic progestogens as well as a variety of other bioactive agents have been one of the primary devices developed over the past one and a half decades. Clinical testing of these devices has continued over much of the period until, at this time, the World Health Organization is planning scaled-up automated production of these devices for very large-scale clinical testing and evaluation of these devices throughout their world-wide system of cooperating clinics.

Controlled release vaginal rings have a number of advantages over intrauterine and intracervical devices and also oral contraceptives. Some of these advantages are

1. After a brief training session, these vaginal devices can be handled (inserted and removed) by the users. Although this requires a higher degree of patient ''compliance'' than that required by physician-inserted intrauterine devices (IUDs) and intracervical devices (ICDs), it is much less compliance than required by a daily contraceptive dosage.
2. They can be easily removed by the user, if desired, and as a consequence can be washed and cleansed in approved manners before reinsertion.
3. They do not necessarily have to be in place during intercourse as long as the progestogen level in the patient does not drop below a certain specific level (i.e., brief removal of the device will likely not influence its efficacy).
4. They can readily be designed to release effective levels of the active agent for long periods (many months or even years), thus not requiring the frequent return of the user to a clinic or pharmacy (also reduced costs)

Vaginal devices were designed and fabricated not only for humans, but initially also for monkeys, in which some early studies were conducted. These devices were generally designed to release in the animal not only the same amount of drug on a body weight basis as the human devices, but a second set of devices was prepared, wherever possible, to release a multiple dose (frequently ten times the human dose) to the animal.

A. Design Considerations

In early experimentation it was found that the progestogens could be released in the desired ranges from appropriately designed and sized silicone polymer devices. Silicones had already been used in a variety of implants and other applications in the body and therefore had a record of biocompatibility as well as the biostability required in this type of application.[1] It was anticipated that, in some instances, these devices might be left in the vagina for months between removals.

It was also noted in release experiments with these silicone rubber devices that a relatively wide range of permeabilities was obtained between progesterone and various synthetic progestogens, e.g., norethindrone and levonorgestrel. Although the total chemical structures of these steroids vary only slightly, it was found that under similar conditions, the permeability of progesterone in cross-linked dimethylsiloxane polymers (Dow Corning Medical Grade Silastic #382) was about 20 times that of levonorgestrel, whereas the permeability of norethindrone was about twice that of levonorgestrel.[2] Another interesting aspect of the relationship between structure and permeability was noted between the two stereoisomeric forms of norgestrel. It was found that when equal quantities of a racemic mixture of the dextro- and levo- forms and of the levo- form alone of norgestrel were placed in similar release devices, the release rate of levonorgestrel was essentially the same from both devices, the dextro- and levo- forms apparently diffusing at their own independent rates with little interaction.[3]

Other practical design considerations have to do with the quantities of levels at which bioactive substances can be mixed with silicone rubber or other polymeric materials and still maintain reasonable polymer properties for subsequent molding or extrusion processing. For example, it was noted that when more than 40% progesterone was added to an uncured silicone rubber, the viscosity of the mixture made it difficult to process, cure was inhibited to the extent that much larger quantities of catalyst were required, and the properties of the system were so poor that subsequent handling of molded parts became very difficult.[4]

The release of a bioactive agent from a polymeric delivery system can be conceptualized in the following steps if the agent (drug) is dispersed in a polymeric or other matrix and surrounded by a polymeric membrane which acts as a semipermeable barrier for release of the drug:

1. Diffusion of the agent in the matrix to the inner surface of the membrane barrier
2. Dissolution in the polymer barrier
3. Diffusion across the membrane based on the concentration gradient developed
4. Desorption from the membrane at the outer surface
5. Dissolution or interaction of the agent with the medium in contact with the external membrane surface
6. Diffusion away from the membrane surface

As this mechanism occurs, the slowest step in this series will dictate the overall release rate and, if this step is held constant (assuming an infinite drug reservoir), a linear or zero-order release may be achieved. Linear release over relatively long periods has been a desirable goal for many clinical applications. Such zero-order delivery of agents over long periods (months or longer) has been achieved in some devices when a small release (i.e., microgram quantities per day) is acceptable from relatively large devices (e.g., vaginal rings) and where a barrier membrane controls the rate of release.

A number of designs for vaginal rings have been developed. A design that has received significant development is the solid suspension of steroid in a permeable polymer. This can be prepared with room-temperature-vulcanizing (RTV) silicone rubber by levigating the steroid crystals in the liquid polymer as the cross-linking catalyst (stannous octoate) is added, after which the polymeric mixture quickly cures to a rubbery state. It can be similarly accomplished using polyethylene as the matrix material and adding the steroid to powdered polymer in a mixing extruder, which mixes and keeps the steroid uniformly suspended as the mixture is returned to room temperature.[5]

Such devices have an advantage in mechanical strength over polymeric capsules; however, if diffusion through the polymer is the rate-controlling step, the release rate may not be very constant with time. This inconstancy of rate is due to gradual depletion of the suspended steroid near the boundaries of the device and the increased time required for the remaining steroid to traverse this depletion zone. This depletion effect can be modified by increasing the amount of crystalline drug within the device, which effectively reduces the diffusion distance at a given point in time with a relatively minimal increase in the measured release rate. It has been shown that when a substance is suspended in such a matrix, and when diffusion through the matrix is the rate-controlling step, the release rate is expressed by $dQ/dt = (ADCs/2t)^{1/2}$.[6] In this expression, Q is the amount of substance released, A is its loading in the matrix (i.e., the amount per unit weight of matrix), DC_s is its diffusion constant and its solubility in the matrix, and t is time. The hypothesis that diffusion through the matrix is the rate-determining step can be tested by determining, after a gradient has been established, whether the amount of substance released plotted vs. the square root of time results in a linear relationship. If, on the other hand, the rate of transfer of substance across the matrix-solution interface controls the rate of release (partition controlled), then the amount released, plotted against time, gives a straight line.[7-9]

Determination of the release rate of medroxyprogesterone acetate into distilled water from a solid suspension of the steroid in silicone rubber did not show the "square root" relationship.[10] In this case, it was found that the second mechanism, which allowed for the slow diffusion of steroid from the boundary of the device, provided a better correlation with the experimental results. Thus it appears that more than one mechanism may frequently be in operation in the release of drugs from polymeric systems. Once the approximate solubility in and diffusion rate through a polymer for a given drug are known, it is possible to design devices that release prespecified daily amounts of a given steroid. This is accomplished by making a steroid-loaded core of the proper dimensions and covering it with the requisite thickness of pure polymer. By using a high concentration of steroid in the core, the release rate will be very close to zero order, at least while the first 30 to 40% of the steroid is diffusing from the device. In systems that utilize relatively low daily doses of some of the very effective synthetic progestogens, devices with the potential for lasting at least 3 years are being tested.

A number of designs for vaginal devices that have been and are being tested are represented schematically in Figure 1. Design A is a homogeneously impregnated device, extensively tested with steroids such as norethindrone, R-2323, and norgestrel.[11-13] Designs B and C are modifications of A, featuring an outer sheath of pure silicone rubber, through which the steroid must diffuse.[14,15] As stated in Fick's law, the release rate from design C is necessarily less than the release rate from design B for a given steroid. Design D is a modification for analytical purposes, in that it permits the relatively high release of drug for a limited period of time and, following removal from the subject, the steroid remaining in the device (and, by difference, the amount released in vivo) can be determined. The design represented by B and C has been fabricated both by a number of molding steps[15] and by using silicone tubing as the outer layer.[16] As an indication of the release rates possible from these types of devices, design D, with an 8-mm outer drug core diameter, will release approximately 4000 μg/day of progesterone;[15] design C, with a 2-mm outer drug core diameter, will release about 20 μg/day of levonorgestrel.[17] Devices E and F were designed to release a single or two different drugs from the same device.[18]

If a drug-loaded polymer matrix is coated with a second polymer that is less permeable to the steroid than the polymer matrix material, this second polymer barrier may then be the rate-determining feature. With diffusion of steroid to the surface no longer the determining

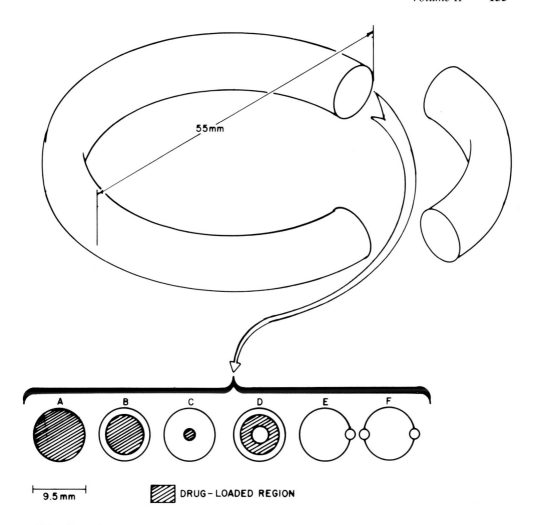

FIGURE 1. Typical designs for vaginal devices: (A) homogeneously impregnated device; (B and C) modification of A, with outer silicone rubber sheath as diffusion barrier for bioactive agent; (D) modification for analysis of amount of released agent; and (E and F) other models for release of one drug or two different drugs.

step, the device will release at an apparent zero-order rate for as long as the steroid concentration at the interface between the barrier membrane and the drug-laden matrix remains sufficiently high to maintain an approximately constant gradient across the barrier polymer — perhaps until 40 to 50% or more of the steroid has been diffused from the device.

B. Molding of Vaginal Rings

The vaginal devices described have generally been molded entirely from RTV silicone materials, although in some instances, where very low doses of some drugs were desired, other polymers have been used to provide rate-controlling barriers to adjust release rate.

The molds used were usually machined from plastics, although in some instances metal molds were also fabricated. The most common mold material has been preshrunk polymethylmethacrylate and, in some instances, polycarbonate. One type of mold providing the toroidal-shaped rings is made in two concentric halves and is held together in the center by a large bolt/nut combination (see Figure 2).

The vaginal rings as shown in designs B, C, and D (see Figure 1) are made using either two or three sets of molds. The rings are molded in the following manner:

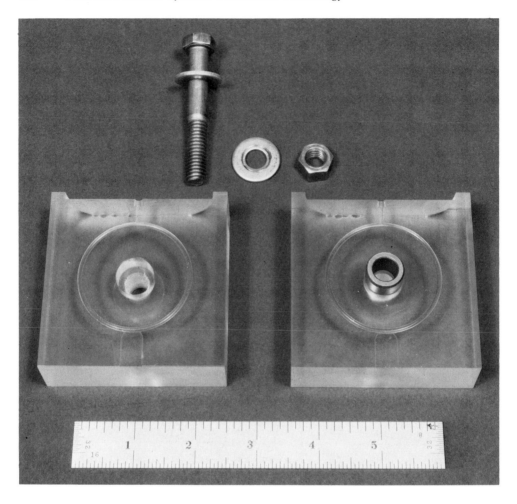

FIGURE 2.   Polymethylmethacrylate mold for a small, drug-laden, core ring (2-mm cross section). At the top of each mold-half is the injection port, while the bottom relief in each half permits an escape from the mold during filling (scale, inches).

1. The drug to be used either in powder form or as a liquid is carefully weighed and thoroughly mixed with a weighed portion of liquid RTV silicone rubber prepolymer to provide a desired concentration of uncatalyzed liquid resin as a master batch.
2. An aliquot of the master batch sufficient to fill a specific number of core molds is withdrawn and catalyzed with an appropriate quantity of catalyst.
3. After thorough mixing, the catalyzed, drug-loaded resin is degassed for 1.5 to 2 min at approximately 29 in. of mercury.
4. The catalyzed, degassed batch is then carefully placed in a thick-walled cylinder (either plastic or steel; approximately 60 to 70 cm^3 working volume). This cylinder has a small hole in its bottom (approximately 2.4 mm). The piston, a close-fitting nylon rod, is placed in the cylinder and the piston/cylinder is inverted and entrapped air is bled out of the cylinder hole until only drug-loaded resin is being extruded.
5. The cylinder is then placed on top of the core ring mold (see Figure 2 — the molds have aligning projections on them to position the cylinder) and, using a press or other leveraging device, the piston is forced into the cylinder to cause resin to flow from the cylinder into the mold through the mated fill port. The pressure is released when

the resin is observed to fill the mold and flushes into a bleed area machined into the bottom of the mold. A number of molds can thus be filled from a single charge of the cylinder.

6. The mold is then placed in a 60 to 70°C oven for 10 to 15 min (cure time can readily be accelerated or slowed by adjusting the temperature and/or quantity of catalyst).

7. After cure, the mold is cooled, the center bolt is removed, and the mold halves are carefully separated so that the cured core ring remains in one half of the mold. Any flash around the ring is removed.

8. The core ring in the half-ring mold is then mated with a full-size ring mold (all sizes of half-ring molds were machined to mate with all others) and bolted together.

9. The cylinder is now filled with pure, catalyzed, degassed Silastic 382, placed on the mated mold halves, and the large half-ring mold is now carefully filled with resin so that the pure resin now covers half the core ring.

10. This mold system is cured in the 60 to 70°C oven for approximately 10 to 12 min and, after cooling, the mold halves are separated with the drug-loaded core ring centered in the large half-ring and both are retained in the large-ring half-mold.

11. This half-mold is now mated with another large half-mold and the processes of points 9 and 10 are repeated to obtain the complete, cured ring with a very well-centered drug-loaded core.

12. The rings are deflashed, if necessary, and washed, packaged, and sterilized before use.

A ring such as D (see Figure 1) which was designed to give a prescribed release rate and also limit the contained drug was made in a manner similar to that described above. However, it was necessary to have three different-sized mold sets, and the finished ring required five molding steps rather than the three described above. The rings and molds were designed not only to deliver a specific daily amount of drug precisely, but also make possible the accurate analytical determination of the exact amount of drug released by the ring over its period of use.

C. Vaginal Rings for Primates

This same basic design was used to prepare vaginal rings not only for humans, but also, on much smaller scale, for testing in primates (rhesus monkeys). These small rings which were 22 mm in outside diameter and 3.9 mm in cross section were designed to release a wide range of progestogens (progesterone, norethindrone, and levonorgestrel). Because both small and large daily doses had to be delivered from these primate rings (i.e., a 1920-μg/day ring and also a 190-μg/day progesterone ring from the same size mold), it was necessary in some cases to coat the inner (or drug-laden core) ring with a thin coat of less permeable polymer such as ethylene vinyl acetate or polyethylene to give very low release rates. In some cases, for maximum delivery, a thin covering (approximately 0.1 mm) of pure silicone rubber was coated over a full-size ring which had been homogeneously loaded (see Figure 1A) with the desired drug at relatively high concentrations.

D. Other Vaginal Rings

Rings were prepared containing other bioactive agents in addition to progestogens. These agents could be generally classified as spermicidal agents. Some agents for which rings were developed were nonoxynol-9, quinine (both as quinine sulfate and as the base), and urea. Due to excessive osmotic swelling, the urea rings were not developed beyond the initial laboratory design. However, both nonoxynol-9 and quinine devices were developed and delivered for clinical testing. The fabrication of these devices was similar to that of the progestogen-releasing devices except for variation in core ring size.

Although the development of automated production of progestogen-containing vaginal rings of this type is well underway, these rings are still being fabricated in Mexico by the thousands by the technique described above for continuing clinical trials. Additionally, a variety of drug delivery systems have been designed using biodegradable polymers in which the drug is released, usually both by diffusion and through polymer matrix erosion. A summary describing some biodegradable delivery systems is given by Skiens et al.[19] and it appears that this type of controlled release device will find increasing use and acceptance.

III. CERVICAL DEVICES (ICDs)

In addition to the development of controlled release vaginal devices, an extensive program was also directed at the development of controlled release devices capable of delivering drugs to the endocervix. The basic assumptions supporting this approach are

1. Such a delivery system could release a sufficiently high concentration of progestogens to the endocervical epithelium, altering cervical mucus properties and inhibiting sperm penetration through the cervix.
2. Locally released progestogens may directly inhibit sperm mobility.
3. The systemic absorption of progestogen thus released would be minimal and would produce no metabolic alterations.
4. Hypothalamic, pituitary, ovarian, and endometrial suppression should be negligible and, therefore, the menstrual cycle should remain undisturbed.
5. Neither the devices nor the progestogen would produce untoward local toxic effects.

Cervical epithelium is highly sensitive to hormonal steroids. Progestogens such as chlormadinone acetate or megestrol acetate orally administered have an inhibitory effect on cervical mucus and sperm migration and are effective contraceptives. These compounds, however, even at the low dosages used, suppress hypothalamic-pituitary-ovarian function and are associated with disturbances of the menstrual cycle, such as breakthrough bleeding, amenorrhea, and prolonged bleeding episodes.[20-22]

Progestogens, such as norethindrone and norgestrel, are rapidly absorbed by the cervix.[22] Application of progestogens directly to cervical epithelium has been hampered until recently by the absence of a device which could be retained for a reasonable length of time and which could release, chronically, a progestational steroid. Cohen et al.[23] showed that the presence of an ICD containing progesterone was effective in inhibiting the normal preovulatory cervical mucorrhea and produced negative postcoital tests. Glass and Morris[24] found that silicone rubber devices releasing chlormadinone acetate interfered with fertilization, implantation, and sperm transport when placed in the cervixes of rabbits. These devices, however, caused enlargement of the cervix in the rabbit and produced stromal and intercellular degeneration and a blunting of the normally columnar epithelium.[25] Preliminary experiments in women with three types of chlormadinone-releasing devices showed effects on the cervical mucus when the estimated release rate was only 0.15 mg/day. With a release of approximately 1.6 to 2.1 mg/day, there were systemic effects, including elevation of basal body temperature, endometrial changes, and uterine bleeding.[23] Together, these preliminary studies suggested that the cervix is capable of retaining a steroid-releasing device that induces alterations in cervical mucus properties and sperm penetration.

Progestogens may also directly affect sperm. Kesseru et al.[26] found that the presence of progestogens in the cervical mucus suppresses and arrests sperm penetration in vitro.

A. Design

A number of devices have been developed to deliver controlled doses of drugs to the

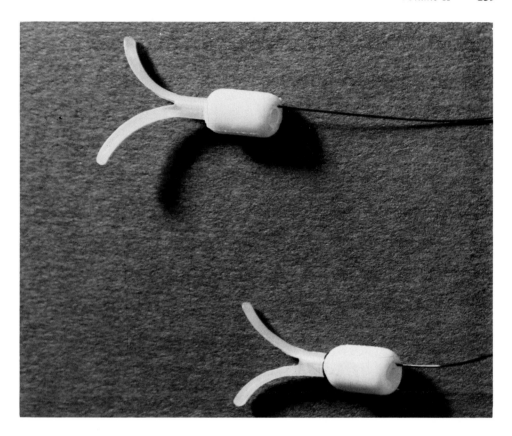

FIGURE 3. Intracervical devices with drug reservoirs attached, developed to study retention characteristics in the cervix. Upper device has arms 14 mm long; lower device has arms 12 mm long. A third device (not shown) has two additional short arms (approximately 4 mm long) at right angle to the 12-mm arms.

endocervix. These include tubes and rods prepared from silicones and containing progestogens for use in rabbits and specially designed devices molded from polypropylene and silicone rubber to deliver drugs to nonhuman primates *(Erythrocebas patas* monkey).[27] Still others were designed for human use in clinical trials. One design of human device with some variations is shown in Figure 3. The primate devices produced were similarly shaped, scaled-down versions of the human device since *E. patas* has a relatively straight cervix and reproductive tract, similar to that in humans.

A variety of designs closely related to the devices shown in Figure 3 have been developed with differing internal configurations for studying the delivery of progestogens and spermicides. Some of these devices are shown in Figure 4. Each of these devices has a central polypropylene tube supporting a modified cylinder of silicone rubber which acts both as the drug reservoir and as the rate-controlling device for drug delivery. Thus, the device differs both in the design of the arms which retain the device in the cervix and in the body which retains the drug reservoir/delivery system. These devices are designed to remain in the cervix for the lifetime of the device. Insertion and removal are carried out by medical personnel. Initial results of a multicenter clinical trial, carried out under the direction of the World Health Organization,[28] indicate an expulsion rate of 27.8% with prototype I (see Figure 3). There were statistically significant improvements in the retention of prototypes II and III, compared to prototype I. Women having retention failures tended to be younger and had had fewer viable births than the other women, in addition to the fact that (perhaps) the 6-mm-diameter reservoir might be too large for this group. The retention rates of prototypes

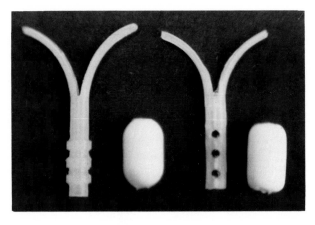

FIGURE 4.   Internal configuration of the intracervical devices. The device with ridges on the left is designed to release progestogens; the device with fenestrations on the right is designed to release spermicides.[27]

II and III were initially considered sufficiently high to proceed with clinical studies of drug-loaded devices for periods of up to 1 year. These studies were later postponed because of other clinical problems with the device.

The ICD on the left of Figure 4 releases microgram quantities of levonorgestrel.[27] This progestogen is ideal for release from the rather limited-size reservoir incorporated into the device since it is a compound which is highly active and thus requires a relatively small delivery level. This device should be able to release norgestrel at levels of 20 to 25 μg/day for 4 years. A variation of this design could potentially reduce the release rate to as low at 1 μg/day, with a correspondingly lengthened total period of release. A limited clinical study using this device was to be carried out under the direction of the World Health Organization.[24]

An ICD was also designed to release the spermicide quinine. The internal design of this device (shown on the right in Figure 4) differs somewhat from the progestogen-releasing device. The progestogen-releasing device (on the left) has ridges on the polypropylene tube to help retain the reservoir, while the quinine-releasing device (on the right) has fenestrations in the tube to help retain the reservoir and to permit spermicide to diffuse to the center of the tube. A multicenter clinical study of intracervical devices that release 20 μg/day of quinine sulfate[28] has shown inhibition of sperm migration in more than 80% of postcoital trials. No adverse side effects, either local or systemic, were noted.

Additionally, devices have been designed to release other spermicides. These include, in addition to quinine, nonoxynol-9 and emetine. These devices (cylindrical rods, 1.5 mm in diameter by 15 mm long) were molded from silicone rubber containing the appropriate spermicide. These rods used in vivo in rabbits were found to release approximately 5.5 μg/day quinine, 200 μg/day nonoxynol-9, and 450 μg/day emetine. The devices were placed into the rabbit's cervix and held in place by a single suture placed immediately above the internal cervical os. Larger doses were obtained by placing as many as three high-dose devices in each rabbit.

B. Fabrication

The fabrication of the intracervical devices shown in the figures was carried out as follows:

1.   The body was molded from polypropylene in a single-cavity aluminum mold base using a small bench-size injection molding machine at 260 ± 3° C and approximately 20,000 psi molding pressure.
2.   Any flash on the polypropylene body was removed.

3. The silicone rubber drug reservoir was prepared by molding in a five-cavity polymethylmethacrylate mold. The material was preblended by mixing the silicone rubber (Dow Corning 382 Medical Grade Elastomer) with 5% barium sulfate (to render the device radio-opaque) and the appropriate amount of drug (progestogen or spermicide). The material was made up as a master batch sufficient to prepare the entire lot of reservoirs desired for the study.

4. For molding of reservoirs, a sufficient quantity of the material was weighed (approximately 3 g) and placed in a container for mixing and approximately 50 mg of the catalyst (stannous octoate) was added and mixed thoroughly.

5. The mixture was degassed in a vacuum system for approximately 90 sec at greater than 29 in. of mercury.

6. The material was then placed in a plastic syringe and each of the five cavities, prefilled with a polypropylene body, was filled through a fill port by extruding the catalyzed silicone rubber/drug blend from the syringe.

7. After filling the cavities around each polypropylene body, the filled mold is placed in an oven at approximately 65°C for 15 min to cure.

8. The ICDs were then removed from the molds and deflashed, if necessary.

9. A monofilament was attached to the body of the ICD just above the silicone reservoir with the monofilament end passing between the body and the reservoir and extending below for an appropriate distance.

10. Fabrication of an inserter tube was carried out by heating a section of polypropylene tubing (4-mm outside diameter, 0.5-mm wall) to approximately 180° C with hot air and shrinking it over a 1.3-mm tapered mandrel. The mandrel was removed and the necked-down portion of the tubing was then cut to an appropriate length (14 mm above shoulder).

11. An inserter rod was fabricated by heating a 1-cm section of a 22-cm-long polypropylene rod (2.2-mm diameter) with hot air from a gun and grasping the heated portion with pliers to neck the hot portion of the rod down to an approximately 0.9-mm section. This necked-down portion of the rod was then cut to an appropriate length (35 mm).

The device system was then washed, dried, and assembled with the exception of placing the gelatin half-capsule over the device arms (this is done just prior to insertion by the physician). The entire device was then placed in a plastic package and, after sealing, the package was sterilized and labeled. Figure 5 shows the complete device including the inserter system.

The two devices shown in the figure are the same device shown in two configurations. In both cases, the device is shown attached to the inserter; however, the bottom device in the figure is shown ready for insertion. This device is sitting on the inserter tube with the wing deployment rod (push rod) inside the tube and the arms held in a collapsed position by means of gelatin retaining capsule placed over their tips. After insertion and proper placement of the device in the cervix, the push rod was moved upward through the tube and device, forcing the gelatin capsule off the device arms into the uterus (where it dissolves) and permitting the device arms to deploy as shown in the device at the top of the figure. This deployment locks the device in place, with the drug-releasing silicone core resting in the upper cervix. The inserter tube and rod are then removed. The black thread attached to the devices is used to determine the position of the device and also to facilitate removal of the device at some later time.

These devices were designed to remain in the cervix continuously, through the menstrual cycle and for prolonged periods. In contrast to the placement of a vaginal device by the user (after initial instruction), the insertion and removal of the ICDs are performed by medical personnel. Further studies of the acceptable retention, efficacy rates, and long-term safety remain to be established for these devices.

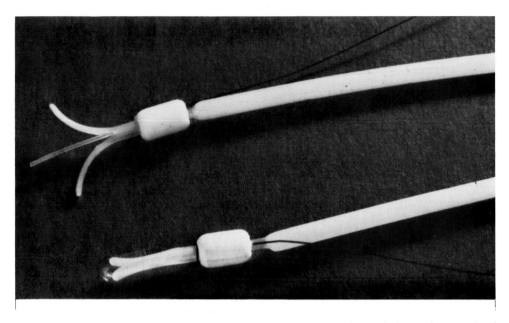

FIGURE 5.   Controlled release intracervical device and inserter system. Device on the bottom is mounted and ready to be inserted. Device on the top represents a device already inserted and with retaining arms deployed, but before removal of inserter system.

IV. CONCLUSION

The value of controlled release systems and their feasibility for fertility control are well demonstrated by both animal and clinical studies. The relatively low doses of contraceptive agents required in an effective delivery system in the reproductive tract is a very desirable feature. Minimal dependence on patient compliance and skills as well as reestablishment of normal fertility patterns after cessation of treatment are also favorable virtues of these systems.

Additional work must be directed at extending the useful life of the devices and the retention time of the active device in body cavities. Acceptable retention in body cavities depends both on geometric shape and surface properties of the device. Also, improved release patterns and control of exact rates will undoubtedly result from further investigation of polymer properties and new methods of placing the bioactive agents in the polymeric matrices. Perfection of controlled release systems will result in additional methods of safe, effective targeted delivery of bioactive agents in the body.

REFERENCES

1. **Dziuk, P. J. and Cook, B.,** Passage of steroids through silicone rubber, *Endocrinology,* 78, 208, 1966.
2. **Duncan, G. W., Burton, F. G., and Skiens, W. E.,** Vaginal, cervical and subdermal delivery systems, in *Human Reproduction: Conception and Contraception,* Hafez, E. S. E., Ed., Harper & Row, Hagerstown, Md., 1979, chap. 34.
3. **Burton, F. G.and Skiens, W. E.,** private communication to World Health Organization, Geneva, 1981.
4. **Skiens, W. E., Burton, F. G., and Duncan, G. W.,** Development of a Polymeric Vaginal Device to Release Spermatocidal Agents for Use in Human, Rep. No. 74163-R to World Health Organization, Geneva, Battelle-Northwest, Richland, Wash., 1982.

5. **Kalkwarf, D. R., Sikov, M. R., Smith, L., and Gordon, N. R.,** Release of progesterone from polyethylene devices in vitro and in experimental animals, *Contraception,* 6, 423, 1972.
6. **Higuchi, W. I.,** Mechanism of sustained-action medication: theoretical analysis of rate of release of solid drugs dispersed in solid matrices, *J. Pharm. Sci.,* 52, 1145, 1963.
7. **Chien, Y. W.,** Thermodynamics of controlled drug release from polymeric delivery devices, in *Controlled Release Polymeric Formulations,* Paul, D. R. and Frank, F. W., Eds., American Chemical Society, Washington, D.C., 1976, chap. 7.
8. **Chien, Y. W. and Lambert, H. J.,** Controlled drug release from polymeric delivery devices. II. Differentiation between partition-controlled and matrix-controlled drug release mechanisms, *J. Pharm. Sci.,* 63, 515, 1974.
9. **Roseman, T. J.,** Release of steroids from a silicone polymer, *J. Pharm. Sci.,* 61, 46, 1972.
10. **Roseman, T. J. and Higuchi, W. I.,** Release of medroxyprogesterone acetate from a silicone polymer, *J. Pharm. Sci.,* 59, 353, 1970.
11. **Mischell, D. R., Lumkin, M., and Jackanicz, T.,** Initial clinical studies of intravaginal rings containing norethindrone and norgestrel, *Contraception,* 12, 253, 1975.
12. **Victor, A., Edquist, L., Lindberg, P., Elmansson, K., and Johansson, E. D. B.,** Peripheral plasma levels of d-norgestrel in women after oral administration of d-norgestrel and when using intravaginal rings impregnated with d,1 norgestrel, *Contraception,* 12, 261, 1975.
13. **Viinikka, L., Victor, A., Janne, O., and Raynaud, J.,** The plasma concentration of a synthetic progestin, R2323, released from poly-silastic vaginal rings, *Contraception,* 12, 309, 1975.
14. **Akinla, O., Lahteenmaki, P., and Jackanicz, T. M.,** Intravaginal contraception with the synthetic progestogen, R2323, *Contraception* 14, 671, 1976.
15. **Burton, F. G., Skiens, W. E., Gordon, N. R., Veal, J. T., Kalkwarf, D. R., and Duncan, G. W.,** Fabrication and testing of vaginal contraceptive devices for release of prespecified dose levels of steroids, *Contraception,* 17, 221, 1978.
16. **Victor, A. and Johansson, E. D. B.,** Plasma levels of d-norgestrel and ovarian function in women using intravaginal rings impregnated with D,L-norgestrel for several cycles, *Contraception,* 14, 215, 1976.
17. **Burton, F. G., Skiens, W. E., and Duncan, G. W.,** Low-level progestogen releasing vaginal contraceptive devices, *Contraception,* 19, 507, 1979.
18. **Schopflin, G., Laudahn, G., Muhe, B., Hartmann, H., and Windt, F.,** Vaginal Rings, U.S. Patent 4,012,496, 1977.
19. **Skiens, W. E., Burton, F. G., and Duncan, G. W.,** Biodegradable delivery systems, in *Human Reproduction: Conception and Contraception,* Hafez, E. S. E., Ed., Harper & Row, Hagerstown, Md., 1979, chap. 35.
20. **Board, J. A.,** Continuous norethindrone 0.35 mg, as an oral contraceptive agent, *Am. J. Obstet. Gynecol.,* 109, 537, 1971.
21. **Moghissi, K. S. and Marks, C.,** Effects of microdose norgestrel on endogenous gonadotropic and steroid hormones, cervical mucus properties, vaginal cytology, and endometrium, *Fertil. Steril.,* 22, 424, 1971.
22. **Moghissi, K. S., Syner, F. N., and McBride, L. C.,** Contraceptive mechanism of microdose norethindrone, *Obstet. Gynecol.,* 42, 585, 1973.
23. **Cohen, M. R., Pandya, G. N., and Scommegna, A.,** The effects of an intracervical steroid-releasing device on the cervical mucus, *Fertil. Steril.,* 21, 715, 1970.
24. **Glass, R. H. and Morris, J. M.,** Antifertility effects of an intracervical progestational device, *Biol. Reprod.,* 7, 160, 1972.
25. **Greenhill, A. H. and Glass, R. H.,** Histologic effects of intracervical chlormadinone impregnated silastic devices in rabbits, *Contraception,* 6, 287, 1972.
26. **Kesseru, E., Camacho-Ortega, P., Laudahn, G., and Schopflin, G.,** In vitro action of progestogens on sperm migration in human cervical mucus, *Fertil. Steril.,* 26, 57, 1975.
27. **Moghissi, K. S., Burton, F. G., Skiens, W. E., Leiniger, R. I., Sikov, M. R., Duncan, G. W., and Smith, L.,** An intracervical contraceptive device, in Drug Delivery Systems, Gabelnick, H. L., Ed., USDHEW Publ. No. (NIH) 77-1238, U.S. Department of Health, Education and Welfare, Washington, D.C., 1977, 79.
28. **Anon.,** Intracervical devices, in Special Programme of Research Development and Research Training in Human Reproduction, 7th Annu. Rep. World Health Organization, Geneva, 1979, 79.

Chapter 8

SELECTIVE DRUG TARGETING

Dean S. T. Hsieh, Diana Chow, and J. Chen

TABLE OF CONTENTS

I. MONOCLONAL ANTIBODY-DRUG CONJUGATES

A. Monoclonal Antibodies

Antibodies are protein molecules produced by vertebrates to combat foreign substances such as bacteria, parasites, and viruses. Each class of foreign substance represents a type of antigen. Each antigen type is neutralized or eliminated from the body by a unique type of antibody. It has long been obvious to the pharmaceutical community that the development of synthetic antibodies would be a major technological breakthrough in drug therapy. Such a production would be a two-step process: (1) isolate an antibody and (2) reproduce and preserve that antibody in large quantities.

Antibodies of a class specific to a given substance have a common origin. They developed in vivo in reaction to the substance itself (an immunizing agent). Because one immunizing agent may contain several antigens, there may exist several antibodies specific to one immunizing agent. Because of the difficulty in isolating an antibody which matches a single antigenic determinant, conventional approaches have utilized mixtures of antibodies in antiserums. To isolate and reproduce a single antibody would be to create a monoclonal antibody, one which would be attracted to a single antigenic determinant.

A method for accomplishing the isolation and reproduction of antibodies has been developed by Milstein.[1] An immunizing agent carrying several antigenic determinants is injected into the body of an animal. The response of the immune system produces distinct types of lymphocytes. Each lymphocyte secretes an antibody (immunoglobin molecule) which matches a specific type of antigen. These lymphocyte cells are then fused with malignant myeloma cells via polyethylene glycol. This fusion enables lymphocytes to rapidly proliferate, producing large amounts of monoclonal antibody.

The implications of this technology are staggering. Artificial antibodies enable scientists to study both the relation between the structure of an antibody and its function and the relation between antibody and antigen during chemical reaction. Practical applications of this technology utilize the natural capacity of the antibody to target antigens in the body. Drugs or radioactive isotopes may be chemically bound to monoclonal antibodies. The bonds will be broken upon contact with the antigen and the drug or isotope will be free to perform its therapeutic task at the site of the affliction.

Monoclonal antibodies may therefore function as transport systems which carry drugs through the body and release them at the site of the affliction they are designed to treat. This type of drug delivery is advantageous for a variety of reasons, including the following:

1. It minimizes the toxic side effects of drugs.
2. It allows drugs to be used in higher concentrations.
3. It allows drugs to be used in smaller quantities.
4. It enhances the ability of a drug to cross membranes.

B. Production of Monoclonal Antibodies

Several methods exist for the production of monoclonal antibodies. Each involves many techniques which may or may not be unique to that method. In general, however, all methods can be conceived as a sequence of five subprocesses:

1. Development of adequate lymphocyte cells
2. Preparation of lymphocyte and myeloma cells
3. Fusion of lymphocyte cells with myeloma cells
4. Selection and isolation of a monoclonal antibody-producing cell
5. Production of monoclonal antibodies

The development of adequate lymphocyte cells may also be conceived as a series of subprocesses which include synthesis of an immunizing agent, immunization of an animal to produce lymphocytes, and the development of myeloma cells. Procedures for producing an immunizing agent will depend on the nature of the chosen antigen. If the antigen is a water-soluble protein, it must be combined with a substance which augments the immune response of the receptor: an adjuvant.[2]

The most commonly used adjuvant was developed by Freund, but alternatives exist. Slavin recommends a 5% aqueous concentration of calcium alginate as an antigen depot.[3] The gel is absorbed slowly, thus requiring fewer inoculations than Freund's adjuvant. The antigen is mixed with sodium alginate and injected peritoneally. Immediately following, calcium chloride is injected into the same site. The calcium alginate which results from the process is absorbed continuously in small doses for 3 weeks or more. Slavin's procedure is followed by intravenous injections which flood the recticuloendothelial system (RES) so that antibody production is increased. A second alternative is proposed by van Preis and Langer.[4] They claim that antigen delivered through a sustained release polymer implant will stimulate the same immune response as an equal dose of antigen emulsified in Freund's adjuvant. Utilization of implant technology as a tool for immunization has the advantage of uniformity, simplicity, and increased choice of delivery sites. It is especially appealing in comparison with Freund's adjuvant for that substance may cause acute inflammatory reactions, the full effects of which are as yet unknown.

Nevertheless, in the typical procedure, antigen and Freund's adjuvant are combined in equal amounts. Two glass syringes are filled, one with the antigen solution in buffered saline or water and the other with adjuvant. They are coupled and locked. The aqueous solution is injected into the adjuvant, and they are passed rapidly back and forth through the syringe. The resulting emulsion is tested by dropping it on the surface of a beaker of water. Although the first drop may spread, subsequent drops should form distinct globules on the surface. The emulsion should appear thick and creamy.[2]

The second subprocess in the development of lymphocyte cells is to administer the immunizing agent to an animal in such a way that the animal's reaction is to form lymphocytes which produce antibodies that match the given antigen. Again, this subprocess may be completed in a variety of ways. The first consideration must be the choice of animal. Mice have been shown to be the most successful recipients because of the availability of mouse myeloma lines.[5] However, in some cases rats may be preferred, e.g., when it is desirable to make an antibody against a mouse protein or because a larger amount of serum is produced per animal.

The site of injection is a subject of debate, but intraperitoneal or intramuscular injections are thought to be most humane. Because soluble monomeric proteins are not normally highly immunogenic, the first injection should be one of the antigen in highly aggregated form. Subsequent injections may be either aggregate or soluble.[6] It is customary to formulate "booster shots" with incomplete Freund's adjuvant to prevent possible hypersensitive reactions. One to three boosters are typical in a period of 2 to 8 weeks. A final injection is typically given 2 to 4 days prior to cell fusion. Lymphocyte cells from both the spleen and the regional draining lymph nodes may be used in fusion.[7]

Several different myeloma lines are now being used to generate hybridomas. However, the only animal for which myeloma cell cultures have achieved a level of practical success is the mouse. A distinguishing characteristic of a myeloma cells is that it secretes large amounts of immunoglobulin of the same idiotype and of the same heavy- and light-chain isotope as the paraprotein in the serum.[8] These are most easily found in mice, but much research is directed toward discovery of myeloma cells in rats and humans.

Fine forceps are used to coax the removed spleen cells through a sterile sieve into a culture medium. This creates a cell suspension. The cells are harvested by centrifugation (400 ×

g for 5 min), twice washed in a culture medium, and counted. Erythrocytes are removed by a diluting aliquots of cells with 3% acetic acid.[9]

There is no well-defined culture medium specialized for preparation of myeloma cells, but most are buffered with HCO_3^-/CO_2. A typical culture medium is 3 to 4 g $NaHCO_3$ per liter in equilibrium with 10% CO_2 in air to give the desired pH. The medium should be kept away from light to avoid the generation of highly toxic photoproducts. Fetal calf serum (10 to 15%) is usually added to the medium, although it is not absolutely necessary to the growth of myeloma cells. Its effect on cell growth is not fully understood, but its use is observed to enhance growth. Cells are seeded at serial tenfold dilutions in petri dishes of medium. The cells grow exponentially. Typical doubling times are 14 to 16 hr. Groups of cells are transferred weekly, from crowded, but not overgrown dishes. If the cells become too dense, they will die.

Several investigators have developed their own methods of fusing lymphocytes with myeloma cells. The following method is that of Galfre and Milstein:[10]

1. Thoroughly mix 10 mℓ of sterile medium (without serum) with 10 g of polyethylene glycol which has been autoclaved and thoroughly cooled.
2. In a sterile 50-mℓ conical-bottom tube, mix 10^8 spleen cells, 10^7 myeloma cells (6 × 10^7 if rat myeloma cells), and serum-free medium to 5 mℓ. Spin at 400 × g for 5 min.
3. Remove all medium by suction; any residual medium will result in the dilution of the final concentration of the polyethylene glycol.
4. Break the newly formed pellet and place it in 200 mℓ of water at 40°C inside a laminar flow hood. Using a 1-mℓ pipette, add 0.8 mℓ of 50% (w/v) prewarmed polyethylene glycol over a period of 1 min, stirring continuously. Continue to stir.
5. Use the same pipette to add 1 mℓ of medium without serum at 37°C; stir for 1 min. Repeat this step three times; the last two times, stir for 90 sec.
6. Add 7 mℓ of warm medium over a period of 2 min, stirring constantly, and then add 12 to 13 mℓ more. Spin at 400 × g for 5 min. Discard the supernatant, break the pellet, and suspend it in 49 mℓ medium with 20% fetal calf serum.
7. Distribute into 48 1-mℓ flat bottom tissue culture wells. Add 1 mℓ of medium with serum and 10^5 normal lymphocyte cells per milliliter of suspension. Store overnight at 37°C in a CO_2 incubator.
8. Remove half the supernatant and add 1 mℓ of a hypoxanthine + aminopterin + thymidine (HAT) medium to each well. Repeat this step daily for 2 to 3 days and then repeat weekly until there is evidence of vigorous growth of hybrids.

A more inclusive view of fusion methods is presented by Goding, but Galfre's method serves as a paradigm example.[11]

Growth of hybrids is another area which leaves little room for debate. Some investigators deposit hybrids directly into HAT medium, while others plate the cells into normal medium for 1 day prior to adding HAT.[1] Normal HT stock solution is prepared by dissolving hypoxanthine and thymidine in 100 mℓ deionized distilled water between 70 and 80°C, sterilizing by filtration, and stored at −20°C in 1-mℓ aliquots. HAT stock solution is prepared by dissolving aminopterin in HT stock solution. Myeloma cells will grow normally in HT medium, but they die within 1 to 3 days in HAT medium.[11] There is no apparent reason for the delay in adding HAT medium, and direct plating into HAT renders it unnecessary to "feed" the cultures for the first 7 days. Subsequent feeding is determined by the rate of cell growth. Feeding is done by removing approximately half of the medium by suction and replacing it with fresh medium. Feeding is done for two major reasons: to gradually dilute out unfused antibodies and to remove waste products while introducing fresh nutrients.

The process by which monoclonal antibodies are produced in mass quantities is called a screening assay. Many hundreds of thousands of such assays must be performed at the initial screening, cloning, expansion and recloning, and bulk culture. Screening assays must be carefully and effectively planned and prepared to ensure efficient and reliable cloning.

The "dynamic range" of a screening assay is the ratio of the strongest signal to the weakest that can be detected above background. The dynamic range of a screening assay will depend on the percentage of the test antigen in the screening substance.[12] For example, if the monoclonal antibody is to be directed against a purified protein, the screening substance is the antigen and the dynamic range ratio will be 1:1. In contrast, if the monoclonal antibody is to be directed against a minor cell surface protein, the required ratio may be as high as 100:1.

A screening assay may be conceived as a series of steps. The first step must induce removal of supernatants. Sterile removal may be accomplished with pipettes. This is an extremely slow process, but alternatives are even more cumbersome and less reliable.[13] Available screening assays include solid-phase radioimmunoassay (soluble protein antigen or cell surface and viral antigens), enzyme-linked immunosorbent assays (ELISA), screening by immunofluorescence, cytotoxic and rosetting assays, and screening by immunoprecipitation and polyacrylamide gel electrophoresis.[13]

Once the growth of hybrid cells is firmly established, cloning should be initiated. Cells may be cloned in soft agar or by limited dilution, which allows for direct testing of the supernatants. Once cloned, the cells must be grown in large numbers for antibody production and frozen for future use.

C. Separation and Purification of Monoclonal Antibodies

Chemically, monoclonal antibodies are immunoglobins which are specific to a particular antigenic determinant. Immunoglobins consist of a group of macromolecules which are highly heterogeneous with respect to charge, mass, and biological activity. This mixture of macromolecules changes with each immunized animal and even with each bleed of the same animal. Nevertheless, the monoclonal antibody produced by an isolated clone is a well-defined chemical rather than a heterogeneous mixture. Its homogeneity simplifies the problem of selecting an isolation procedure for a particular monoclonal antibody from an in vivo or in vitro culture supernatant.

In the last three decades, many different methods for isolating antibodies have been suggested. Conventional procedures for purification of immunoglobins are generally based on the properties of a majority of antibody molecules in a polyclonal mixture. However, each individual monoclonal antibody has distinctive properties, and they are not always those of the majority. Therefore, it is advisable to monitor the purification steps with specific serological assay. A general strategy for purification should utilize procedures which separate molecules on the basis of properties which do not vary widely between different antibodies in the same group, for instance, isolating immunoglobins. The first type of method uses nonspecific means of isolating purified preparations of immunoglobins of various classes. It is often used to form a concentration of immunoglobins suitable for further purification by other methods. The second method is specific. It directs the removal of antibodies from the general mass of serum proteins with the aid of corresponding antigens. Nonspecific methods include

1. Salt precipitation (Kekwick's method)[14] — A crude separation of most species of immunoglobins can be achieved by successive precipitation at different concentrations of saturated ammonium or sodium sulfate at room temperature. The resulting fractions can be further purified by a suitable procedure. All monoclonal immunoglobins, G type (IgGs) can be separated by this technique irrespective of their subclasses.

2.   Ion-exchange chromotography[15] — Proteins become fixed to the ion exchanger through electrostatic bonds and are easily eluted from the absorbent by raising the ionic strength and or by changing the pH toward the isoelectric point of the proteins. Monoclonal antibodies of IgG class are most frequently purified by this method using DEAE-cellulose columns.[16]

3.   Gel filtration — The separation in gel filtration depends on the different abilities of the various sample molecules to enter pores which contain the stationary phase. Very large molecules are unable to enter the pores and so move down the column faster, while the smaller molecules enter the pores and move more slowly. Thus, molecules are eluted in order of decreasing molecular size. This method is particularly suitable for purification of immunoglobin, M type (IgM). Due to its larger size, it passes down a Sephadex® column much faster than other immunoglobins.[17]

4.   Gel electrophoresis[18] — This is a widely used technique for separating charged macromolecules. It is generally run in a 0.8 to 1.0% agarose gel containing pH 8.6 buffer. Most proteins migrate through the gel toward an anode; the extent of the migration is determined by the charge on the protein. High recovery of isolated immunoglobin from serum is obtained by this method.

5.   Isoelectric focusing[19] — This method is a powerful tool for separation and characterization of amphoteric biopolymers because it offers high resolution and reproducibility. Proteins are separated by their isoelectric points in a pH gradient being created by ampholytes.

An example of a specific method is immunoaffinity chromatography.[20] This method exploits the selective affinity of an antigen molecule toward its specific antibody. An antigen is coupled to an insoluble matrix. A serum-containing antibody is added to the antigen matrix. A suitable eluting agent is used to elute the bound antibody. This method of separating and purifying monoclonal antibodies is extremely rapid and convenient in comparison with any of the nonspecific methods. Furthermore, it is independent of both the binding affinity of the monoclonal antibody for the antigen and the availability of the antigen against which the monoclonal antibody is directed.

Difficulties are still encountered in the attempt to purify monoclonal antibodies obtain from in vitro hybridoma cultures. These are predominantly due to the low concentration of supernatants (usually less than 20 g/mℓ) and also to the presence of other serum proteins. Nonspecific methods of immunoglobin purification are not adequate to deal with such cases, whereas immunoaffinity chromatography appears likely to perform the separation efficiently. Nevertheless, a combination of specific and nonspecific methods is commonly utilized to achieve the optimal separation and purification of monoclonal antibodies.

D. Monoclonal Antibody Conjugates

Antibodies have long been identified as potential site-directed drug delivery systems. The discovery and development of monoclonal antibodies have stimulated major research efforts in targeted pharmaceuticals. A prime example and major beneficiary of this research is cancer chemotherapy with regard to tumors. Conventional methods of treating malignant tumors with toxic reagents have experienced the limitations inherent in destroying cancerous cells with drugs which are equally hazardous to normal cells. Pharmacologists thus endeavor to develop carrier systems which safely transport toxic drugs to the desired site in concentrations necessary to elicit the desired therapeutic effect. Important features of these carrier systems include retention of the physical and chemical integrity of the drug in the biological system, ability to cross membranes in order to reach the target site, and selective association of the carrier with the target, followed by the release of the drug moiety at the targeted site. Present limitations to the development of such systems include blood tissue barriers, cell membrane barriers, antigen variants, and tumor cell heterogeneity.[33]

Table 1
MONOCLONAL ANTIBODY-DRUG CONJUGATES

Toxin/cytotoxic agent	Carrier	Type of monoclonal antibody	Reagent	Ref.
Ricin	None	MOPCZl (IgG1)	SPDP	21,22
Methotrexate	None	Anti-BSA (rabbit)	DCC[a] ECDI[b]	23
	Polylysine	None	Acetic anhydride ECDI/N-hydroxysuccinimide	24
	Copolymer: glutamic acid and lysine	Anti-3LL or anti-YAC	ECDI	25
	Dextran T-40	Anti-HLA	NaIO$_4$/Schiff base/ NaBH$_4$	26
Daunomycin	None	Anti-BSA (rabbit)	CNBr	27
	None	Anti-BSA (rabbit)	NaIO$_4$/Schiff base/ NaBH$_4$ Glutaraldehyde ECDI	28
	None	791T/36	Succinic anhydride cis-aconitic anhydride SPDP	29
	Polylysine	None	Succinic anhydride/N-hy-droxysuccinimide/ECDI	24
	Dextran	Anti-YAC (goat)	NaIO$_4$/Schiff base/ NaBH$_4$	25
	Melanotropin	None	NaIO$_4$/Schiff base/ NaBH$_4$	30
	(Fab')$_2$	Anti-BSA (rabbit) or anti-YAC (goat)	NaIO$_4$/Schiff base/ NaBH$_4$	25
p-Phenylenediamine mustard (PDM)	Dextran	Anti-EL4	CNBr/glutaraldehyde (spacer: HMD[c])	31
Mitomycin C	Dextran	H-1 IgG	NaIO$_4$/Schiff base/ NaBH$_4$	32

[a] DCC = dicyclohexyl carbodiimide.
[b] ECDI = 1-ethyl-3-(3'-dimethylaminopropyl) carbodiimide.
[c] HMD = hexamethylene diamine.

Monoclonal antibodies provide a solution to the problem of delivering toxins directly and only to the afflicted area of the body. Table 1 lists examples of monoclonal-drug conjugates reported in the literature. "Immunotoxins" are hybrid molecules: a highly specified antibody is coupled to a toxic moiety (i.e., chemical poison or drug). Examples of toxic moieties are gelonin, abrin, ricin, and diphtheria toxin. Such toxins are composed of two polypeptide chains, A and B. Neither of these two individual chains is highly toxic to cells. The "B" chain binds to terminal galactosyl residues on plasma membranes of most cell types. The "A" chain penetrates the cell membrane and inhibits protein synthesis by enzymatically altering the binding of the 60S microsomal subunit to elongation factor 2. For instance, the heterodimeric glycoprotein ricin is divisible into two distinct subunits or chains. The "A" chain functions as an irreversible inhibitor of eukaryotic cellular protein synthesis (RTA) and the other functions as a lectin with specificity for galactose-containing cell surface macromolecules (RTB).[22]

Coupled with a monoclonal antibody, ricin expresses high toxicity, but the specificity of the hybrid is as broad as the class of galactose-containing cell surface macromolecules. This specificity can be controlled in vitro by blocking the ricin-galactose binding site through the addition of lactose.[34] Hybrid molecules composed of a monoclonal antibody and only the RTA chain are highly specific, but only variably toxic and plagued by difficulties in purification. To overcome problems of purification, pokeweed antiviral protein has been suggested as a substitute for ricin A chain.[35] To overcome problems of variable toxicity, activating agents may increase the cytotoxicity of an antibody-A-chain conjugate. According to one study, NH_4Cl renders an antibody-A-chain conjugate ten times more toxic than intact ricin with lactose.[36] Moreover, B chain plus lactose has also been identified as an A-chain activating agent.[37]

Nevertheless, success with immunotoxins has mainly been achieved in vitro. For example, Cushley describes a treatment for systemic leukemia wherein bone marrow is extracted from the patient prior to chemotherapy and reinfused after tumor cells are destroyed and removed safely in vitro.[38] However, while in vitro studies of immunotoxins become more and more promising, their extension to in vivo applications has met with nonspecific side effects.[39] Limitations exist in selectivity and in the choice of both antibody and toxin.

Although immunotoxins are undoubtedly the major thrust of research in monoclonal antibody-drug conjugates, many other types of these systems exist. Monoclonal antibodies have been successfully linked to toxins, radioisotopes, antimetabolites, alkylating agents, enzymes, and interferon.[40] Monoclonal antibody-drug conjugates may be classified according to the method by which drugs are linked to antibodies: direct linkage or linkage via a dextran (polymer) bridge. Factors which lead to a decision on linking methods include (1) the desired ratio of number of molecules of drug per antibody and (2) variations in the cytotoxicity of drug which are dependent on the binding site.

The desired drug-antibody ratio itself is dependent on several factors. Simplicity of experimental models sometimes suggests a 1:1 ratio. These results may be compared to subsequent models to aid in documentation. Manabe et al. experimented with variations in methotrexate (MTX) antibody ratios, using dextran T-40 as a multivalent carrier:[26] 9.29 mol of MTX were bonded to each mole of anti-HLA IgG_1 (H-1); 90% of control antibody activity was retained and recovery of original protein was 68%. Manabe claims that his method is superior to previous efforts (carbodiimide, mixed anhydride, and active ester) because the dextran bridge will allow several other molecules to be incorporated into the already cytotoxic conjugate.[41] In a study of daunomycin-monoclonal antibody conjugates, Gallego et al. agree that an increase in the number of drug molecules per molecule antibody may be beneficial, but warn that increasing the molecular size of the conjugate may affect the stability of the carrier-antibody bond, the clearance in vivo, and the drug administration route in vivo.[29]

The main issue involved in variable toxicity is that different cell lines may have divergent mechanisms for the internalization of similar cell surface antigens.[21] Variations in cytotoxicity which are dependent on the binding site have been observed in experiments with ricin, with the result that disulfide bonds are preferred in direct linking.[21,42] However, mechanisms of internalization are mostly at the hypothetical stage and not well documented.

Targeted drug carrier systems are only one practical application of monoclonal antibody technology. Another is target cancer radiotherapy. Tradiational cancer radiotherapy involves doses of radiation of 40 to 60 rad/hr. This treatment is plagued by biologically toxic side effects and the destruction of normal tissue. However, the use of monoclonal antibodies conjoined to radioactive isotopes has been demonstrated to decrease the necessary radiation dosage to only 2 to 5 rad/hr. Radiolabeled antibodies are monoclonal antibodies specific to tumor cells combined with a radioactive isotope such as ^{131}I.[43] Iodine may be oxidized and attached by the active set of peroxidase and transferred to tyrosine on the monoclonal antibody.

Upon administration, radiolabeled antibodies absorb neutrons, releasing energy which spans and destroys several cells. Use of ^{10}B as a radioactive agent reduces the effects of the reaction to only 10 μm, or a single cell length. Although the imperfect specificity of antibody

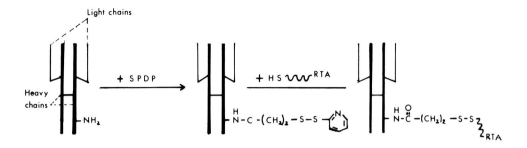

FIGURE 1.   Synthesis of an immunotoxin.

isotope conjuncts renders this technique unavoidably harmful to some normal tissue, this method allows for tumor destruction without significant adverse effects on normal tissue.[43]

Techniques for binding drugs to monoclonal antibodies are dependent on the chemical properties of each component of the conjugate. For instance, monoclonal antibodies may be treated with *N*-succinimidyl-3-(2-pyridyl[dithio] propionate) (SPDP) to produce a pyridyl dithioproponyl-derivatized antibody. In the case of ricin, a disulfide bond is formed upon combination with isolated RTA; the result is an immunotoxin (Figure 1). As an example, human interferon may be coupled to monoclonal antibodies specific to Epstein-Barr virus membrane antigen by the following process:[44]

1.   Combine 6 mg of either substance with 1 mℓ of a stirred phosphate-buffered saline (PBS) solution (pH 7.2).
2.   React with 8 $\mu\ell$ of 20 nM SPDP solution for 20 min at room temperature.
3.   Dialyze interferon for 16 hr against PBS.
4.   Dialyze monoclonal antibody against sodium acetate buffer (pH 4.5).
5.   Incubate with 25 nM dithiothreitol for 15 min.
6.   Dialyze further against PBS.
7.   Mix derivatized IFN with thiolated monoclonal antibody.
8.   Incubate for 2 hr at room temperature and then incubate overnight at 4°C.
9.   Pass IFN-monoclonal antibody conjugate through a 20-mℓ Sephadex® G-200 column to remove uncoupled IFN.

Dextran bridge conjugates follow a quite different coupling procedure. The following is the method of Hurwitz et al. for an anti-Thy-1 antibody-adriamycin conjugate (see Figure 2).[45]

1.   Mix 0.2 mℓ of 7 mg/mℓ carboxymethyl-dextran hydrazide in water with 0.75 mg/mℓ of adriamycin in water.
2.   React overnight at 4°C.
3.   React 4 mℓ solution with 7.5 mg anti-Thy-1 antibody in 4 mℓ of the solution from Step 2 by cross-linking with 0.1% glutaraldehyde for 20 min at room temperature.
4.   Separate on Sephacryl® S-300 for the evaluation of the amount of high molecular weight aggregates (which prove to be negligible).

In-depth studies of preparation techniques for antibody-drug conjugates are covered in other articles and volumes.[46] However, it must be made obvious that applications for monoclonal antibody conjugates are by no means restricted to the treatment of tumors. For example, research is underway in monoclonal antibody conjugates directed against leukemia.[47] The scope of applications for this technology has only begun to be explored.

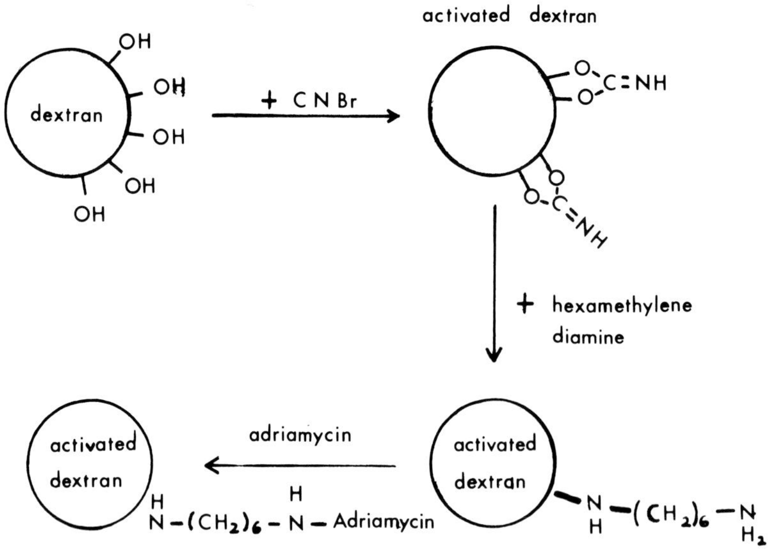

FIGURE 2.   Structure of a dextran-adriamycin conjugate.

II. ALBUMIN MICROSPHERES AS DRUG CARRIERS

Albumin as a drug delivery system seems a natural consequence of the discovery that albumin has a strong tendency to bind drugs. As is often the case, a drug introduced to the body by conventional means may be bound to albumin once in the blood plasma. Blood courses through the body carrying the albumin-bound drug complex; thus albumin is a natural drug carrier. Investigation into the use of human serum albumin microspheres as drug carriers is directed at standardizing that process. Macroaggregates have also been useful in these investigations, but present difficulties in the preparation of successive batches of uniform particle size. Albumin microspheres avoid that problem. They are produced with techniques which involve heated oil emulsions, and it is possible to control the mean diameter of individual particles. As Kramer observes,[48] "Albumin microspheres provide a potentially useful means of delivering drugs to endocytic cells because they are physically and chemically stable, rapidly removed from the vascular system by phagocytosis, amenable to preparation in large batches, nonantigenic, metabolizable, and capable of accommodating a wide variety of drug molecules in a relatively nonspecific fashion." Moreover, microspheres can be selectively entrapped in various tissues by manipulation of microsphere size or mode of administration. Albumin drug carriers are seen as a natural extension of the success radiologists have achieved by using radiolabeled albumin microspheres (aggregates and sulfur colloids), together with the natural phagocytic activity of the liver and spleen to study and diagnose the function of the RES.[49]

Albumin microspheres are produced by two different methods: thermal denaturation and chemical cross-linking.[50] The thermal denaturation method forms microspheres from a water-in-oil emulsion at elevated temperatures. The chemical cross-linking method achieves microsphere production by emulsifying the albumin-drug dispersion in organic solvents and then stabilizing the microsphere by cross-linking with aldehydes at room temperature. The processing parameters of energy input in the initial emulsification step, final temperature level attained, and length in time of the cross-linking reaction may be manipulated to form microspheres of various diameter and "hardness". Scanning electron microscopy and protein staining following by ultramicrotome sectioning show that the microspheres have a porous internal matrix. The degree to which the microspheres are absorbent depends on the temperature at which they are prepared.[51]

Application of albumin microspheres has been mainly concentrated in chemotherapy research. However, researchers in the field have attempted to achieve prolonged insulin release for the treatment of diabetes and to target the lungs for the treatment of emphysema. Most studies have utilized the mouse, rat, and rabbit as the animal model. They have attempted to introduce albumin microspheres in vivo through implantation, intra-arterial injection, and intravenous injection, with subsequent localization by external magnetic fields. It has been documented that both the size of microsphere and the mode of administration play critical roles in the biodistribution of albumin microspheres.[52] The larger albumin microspheres (these range approximately from 12 to 44 μm in diameter) are entrapped in the capillary beds of the alveolar sacs in the lungs, while small microspheres (approximately 1 μm in diameter) filter through the lungs and make their way to the liver. Introduction of any particulate matter in vivo sparks concern of possible toxic effects; yet acute (7-day) and chronic (90-day) evaluation of intravenous injection of albumin microspheres demonstrated negligible adverse side effects and no significant change in histopathological study of vital organs.[53] Furthermore, no adverse effects were generated upon injection of radiolabeled nondrug-carrying human albumin microspheres into a human volunteer.

Research has also been done to demonstrate the effects of concentrating albumin microspheres at one specific site (e.g., in tumor tissue). Small magnetic particles were prepared by phase-separation emulsion polymerization of 25% human serum albumin solution traced with ^{125}I-albumin containing adriamycin HCl and Fe_3O_4.[54] Surface electrostatic repulsion of albumin-magnetic microspheres allows them to avoid the vessel occlusion phenomenon which results from in vivo aggregation of most magnetizable materials. Other advantages of albumin-magnetic microspheres include their ability to carry a wide variety of drugs at predictable release rates, their biodegradability, and their surface properties which permit maximum biocompatibility and minimal antigenicity. External magnets serve to localize albumin-magnetic microspheres at the desired site. Magnetic microspheres will be covered in more detail in the next section of this chapter.

Not only is it possible to incorporate drugs in the body of an albumin microsphere, but researchers have also linked drugs to free amino groups on the surface of albumin microspheres. Martodam et al. employed this technique in targeting inhibitors of leukocyte cells to the lungs.[52] By using microspheres of a relatively large diameter, they ensured that a large percentage were entrapped in the lung vasculature. This study calls attention to the prime concern of researchers in microsphere applications. Though a viable contender for site-specific drug delivery, problems may arise due to chronic entrapment of microspheres in specific areas of the body. There exists the possibility of permanent damage to the blood vessels which feed the alveolar sacs. However, design and development of new applications for albumin microspheres may prove them to be an effective and efficient vehicle for drug delivery.

III. MAGNETIC DRUG DELIVERY SYSTEMS

One type of magnetic drug delivery system has already been mentioned in this chapter: the magnetic albumin microsphere. It has been noted that surface electrostatic repulsion allows albumin microspheres to avoid vessel occlusion. Moreover, these microspheres meet the six major property requirements of an ideal magnetically targeted drug carrier. They are as follows:[55]

1. Small size
2. Adequate magnetic responsiveness
3. Capability of carrying a wide variety of chemotherapeutic agents
4. Predictable release rates of the chemotherapeutic agent

5.   Nonreactive surface properties
6.   Biodegradability

Widder et al. outline two methods of producing albumin-magnetic microspheres: heat stabilization and carbonyl stabilization.[61] Both begin with an aqueous solution of 250 mg human serum albumin, 0.02 ml ^{125}I-bovine serum albumin, 32 mg bulk purified doxorubicin hydrochloride, and 72 mg ferrosoferric oxide. In the heat-stabilization method, a 0.5 aliquot of the solution is added to 30 ml of cottonseed oil to form an emulsion. The emulsion is homogenized by sonication (4°C for 1 min) and then added drop by drop into 100 ml of stirred (1600 rpm) and heated (110 to 165°C) cottonseed oil for 10 min. The stirring is continued as the oil cools to 25°C. The microspheres are washed with anhydrous ether and centrifuged for 15 min at 2000 × g. The last step is repeated three times. The spheres are air dried in the dark at 4°C for at least 25 hr before use. The carbonyl-stabilization method of producing microspheres is very similar to the heat-stabilization method. The above-described aqueous solution is added drop by drop to 100 ml of stirred, but not heated, cottonseed oil for 10 min. The spheres are washed with ether and then resuspended in ether containing either 0.2 M 2,3-butanedione or 1.0 M formaldehyde as a cross-linking agent (40-mg microspheres or 100 ml ether solution). This suspension is stirred rapidly (15 to 60 min). The excess carbonyl reagent is removed by adding 100 ml ether, centrifuging (2000 × g for 10 min), and decanting the supernatant. The last procedure is repeated three times. The microspheres are then lyophilized and stored at 4°C.

Microspheres produced by these two methods have similar characteristics.[57] The size of microsphere is relatively uniform (0.2 to 1.5 μm), as is the quantity of drug released. Atomic absorption spectroscopy was found to have important implications with respect to the magnetic responsiveness of the carrier. The efficiency of drug incorporation may be measured with spectrophotofluorometric analysis. The structural integrity of the albumin matrix before and after application is directly related to the stabilizing treatment. Spectrophotofluorometric analysis demonstrates the presence of doxorubicin HCl when microspheres are incubated in human serum rather than polysorbate 80. Finally, preliminary acute (7 days) and chronic (90 days) toxicity studies with magnetic microspheres in BDF[1] mice found minimal adverse effects.

Limited success with magnetic microspheres suggests that they present a realistic possibility for site-specific drug delivery.[58] Their small and relatively uniform size range permits capillary-level distribution and thus saturation of the desired target site. The magnetic albumin microsphere clearly warrants further investigation as a targeted drug delivery system.

IV. DEXTRAN MICROSPHERES

A traditional method of targeted drug delivery has been the injection of a therapeutic agent into a major artery which leads to the targeted site. Recently, experiments have been aimed not only at directing drugs to a specific site, but retaining the drug at the site once it has arrived. A promising vehicle for accomplishing this aim has been the dextran (or starch) microsphere. Dextran microspheres are made of cross-linked polysaccharide derivatives which are degraded by plasma amylase.[59] The rigidity, size, and degradation rate of the microsphere are dependent on the degree of cross-linking. These variables make it possible to achieve a temporary local arterial flow reduction by interarterial embolization of the microsphere.[60] Blood flow returns to normal upon degradation of the microsphere.[54]

This process allows researchers to deliver a drug to a particular organ and effectively entrap the drug at the targeted site. One study has demonstrated that interarterial injection of dextran microspheres reduced local blood flow in dogs to less than 5% of that in dogs who were not injected with microspheres. Concentrations of drug in areas of the body other than the targeted site were determined relative to the concentration of drug at the targeted site (the kidney cortex). Relative concentrations derived by this method were compared to relative concentrations of the drug injected without microspheres. The concentrations of drug injected with microspheres decreased in the heart, liver, and kidney by factors of 2.8, 3.3, and 3.5%, respectively.[61]

The same mechanism which holds drugs entrapped in a particular organ may also serve to keep oxygen out. By injecting dextran microspheres in the superior mesenteric artery of pigs, Arfors et al. were able to reduce the oxygen pressure in the intestinal mucous membrane to near zero.[59] Because molecular oxygen is known to enhance the effects of ionizing radiations of low linear energies, the advent of dextran microspheres will produce a wave of experimentation in radiation therapy.

Another significant application of dextran microspheres is Wright's research in partial spleen embolization.[60] Studies were designed to determine the optimal embolic agent out of five possible candidates. Materials were compared with regard to ease of use and level, degree, and homogeneity of embolization. It was concluded that dextran microspheres stayed in suspension longer, dispersed more evenly, and produced more homogeneous and peripheral embolization than any other candidate. Dextran microspheres are indeed a unique and widely applicable pharmaceutical tool.

V. pH-SENSITIVE LIPOSOME DELIVERY

When phospholipids are allowed to swell in aqueous media, they form lipid spheres or liposomes.[63,64] Liposomes are either single or multiple aqueous compartments of lipid bilayers. When phospholipids are combined with water (or water plus solute), the result is concentric vesicles of lipid molecules in bimolecular sheets. The water (and solute) enters these vesicles before they completely close, producing liposome-entrapped water (and solute). Sonification will break up these multilayer liposomes into monolayer structures. Their potential as a drug targeting system stems from their capacity to accommodate both aqueous and lipid-soluble substances in their respective aqueous and lipid phases.[65]

Oral administration normally results in the liposome-encapsulated microspheres concentrating mainly in the stomach and intestines, while intraperitoneal injection produces a more even distribution, although approximately one third of the microspheres concentrate in the liver.[66] However, researchers have attempted to target liposome delivery systems by altering such factors as size, charge, fluidity, route of administration, and temperature of target site.[67,68] The most successful application of liposome site-directed delivery has been accomplished by altering the pH of the liposome capsule. This technique involves the use liposomes which contain the pH-sensitive lipid palmitoyl homocysteine. pH-sensitive liposomes are viewed as a technological breakthrough in site-directed drug delivery for several reasons, including the following.[68]

1. The interstitial fluids of most tumors are known to have an ambient pH value which is significantly lower than that of normal tissue. This value may be decreased by the administration of glucose, $NaHCO_3$, or CO_2.
2. Microscopic domains of metastases may also have a lower pH than normal tissue.
3. The exudate pH of an inflammation or infection decreases considerably 60 hr after the start of the reaction.
4. The capillary permeability of both tumors and inflamed sites is greater than that of normal tissue.

Liposomes designed to release drugs when exposed to a low pH have met with success in vitro.[69] Current research is exploring applications in vivo.

VI. REDOX DRUG DELIVERY

The brain presents a unique problem for site-specific drug delivery. It is surrounded by an endothelial capillary wall commonly known as the blood-brain barrier. This barrier is effective in protecting the brain from potentially harmful chemicals, but renders the administration of potentially beneficial drugs to the brain difficult, if not impossible.

Capillaries of the blood-brain barrier discriminate between molecules on the basis of (1) their molecular weight and (2) their lipid solubility.[70] Molecules must pass through these cells to reach the brain. In 1981, Bodor et al. published a response to this quandary: a chemical redox drug delivery system.[70] Drugs are bonded to a quaternary salt which is chemically reduced to a lipoidal dihydropyridine carrier. Upon administration, the compound is distributed throughout the body. The lipid-water partition ratio of the carrier allows it to deliver the drug to the brain. The compound is oxidized in vivo and reverts to its original form. The ionic, hydrophilic salt is quickly eliminated from the body, except that the blood-brain barrier (which works both ways) prevents its elimination from the brain. Enzymes remove the drug from the carrier, providing sustained release of drugs to the brain. The carrier, if properly selected, will then be eliminated from the body.

The Bodor's et al. study in 1981 utilizes phenylethylamine coupled with nicotinic acid. They later extended the redox technology to include much larger molecules — specifically, testosterone. This study represents a potentially important application of steroids in general. Moreover, it asserts that a chemical redox delivery system may also be an appropriate method for delivering drugs to the testes.

VII. POLYMERIC DRUGS

Designers of polymeric drugs utilize techniques of polymer chemistry and affinity chromatography to enhance the effectiveness of existing pharmaceutical agents. Polymeric drugs may be either bioactive polymers or a conjugate consisting of a drug and a polymeric carrier.[72] Polymeric carriers attempt to increase the target specificity, alter the pharmacokinetics, and/or modify the intracellular penetration of the drug to which they are bonded.[73]

The design of a polymer-drug conjugate is a function of the targeted site, the type of affliction, and the type of original drug.[74] Its basic structure is that of a polymer backbone to which drugs and targeting agents are bonded. The polymer backbone has several important functions:[75]

1. To protect the drug molecule from degradation
2. To limit toxicity and antigenicity
3. To provide controlled and prolonged drug delivery through molecular weight selection
4. To provide a matrix for multiple attachments and strong target binding

Targeting agents include antibodies, enzymes, hormones, and compounds with specific endocytotic receptor interactions.[76]

A particular polymer-drug bond is most often chosen on the basis of the targeted site. Permanent covalent attachment of the active ligand is useful only if the conjugate itself is active. Usually a spacer arm is called for to eliminate any steric interference from the polymer backbone. Stable linkages are usually used for conjugates which act at extracellular or pericellular sites. Temporary covalent or ionic attachment is required for conjugates which act intracellularly.[73] High molecular weight polymers (5000 to 10,000) present difficulty for two reasons: (1) they are not absorbed orally and (2) when administered intravenously, distribution and elimination may be retarded due to their large size and physiological limitations. Nevertheless, endocytosis of large polymers occurs in vivo. The rate is dependent upon the cell type, polymer size, and charge; this process is faster for larger, negatively charged molecules.[72]

Polymeric carriers are particularly attractive in their chemotherapeutic applications. The increase in capillary permeability and reduction in lymphatic drainage around cancerous tumors allows a polymeric-drug conjugate to reach the tumor and reside longer in the surrounding extracellular fluid. Moreover, tumor cells exhibit a moderate to marked increase in endocytotic activity. Examples of encouraging results with polymeric drugs include the following:

1. Rowland et al. administered *p*-phenylenediamine mustard (PDM) and specific rabbit antibody (Ig) linked via carbodiimide bonds to polyglutamic acid (PGA) to some mice and administered the Ig separately to others.[77] The greatest increase in survival rate was obtained for the PDM-PGA-Ig conjugate, although there did exist an "antibody-drug effect" when PDM-PGA and Ig were administered separately.
2. Chu and Whiteley coupled methotrexate to albumins and to several dextrans by condensation reactions.[78] They used these to treat mice for leukemia and obtained promising results.
3. Tyrrell and Ryman studies intercalated adriamycin-DNA complexes loaded into "erythrocyte ghosts".[79] These may be effective in treating malignancies of the liver and spleen as well as histiocytic medullary reticulosis.[80]

Although research in polymeric drug delivery appears quite fruitful, currently there are no polymeric drug products on the market or in clinical studies. The major obstacle of such a step has been the rapid clearance of polymer-drug conjugates by the RES.[81] It has further been documented that phagocytosis of polymer conjugates is usually toxic to macrophages and may cause hepatosplenomegaly.[82,83] However, phagocytosis is dependent on the surface characteristics of the particle and macrophage. Hydrophobic interactions, negative charges, and opsonic factors such as IgG and complement factor C_3b have been demonstrated to enhance this process.[84] Furthermore, polyethylene glycol has been used to successfully alter the surface characteristics of certain acrylic carriers and significantly reduce clearance by the RES.[85] Another problem with polymer-drug conjugate implementation has been that large biostable polymer carriers are not metabolized or cleared by the kidney; hence lysosomal overloading may result in bursting of cells and subsequent inflammation.[83] This problem is surmounted by using polymers capable of being broken down into small fragments which are endogenous or easily cleared by the kidney. A second problem with size is that a large molecule often retards diffusion and passage through cell membranes. This prevents the drug from reaching the biophase. Linkage to a polymer has often reduced the solubility and therapeutic effectiveness of a bioactive compound or targeting moiety.[86] Polymer-drug conjugates also have the potential to be immunogenic. Generally, biodegradable polymers are less antigenic than biostable carriers, homopolymers less than copolymers, and antigenic determinants, which are oriented away from the surface, less available to lymphocyte-bound Igs.[87]

The potential rewards of polymeric drug delivery warrant continued research in this field. Success has been achieved in animals for a few drug classes. The production of efficient targeting agents remains a major practical obstacle. Problems associated with carrier linking and site-specific conjuncts must be resolved before the true potential of polymeric drugs may be realized.

REFERENCES

1. **Milstein, C.,** Monoclonal antibodies, *Sci. Am.,* (243)4, 66, 1980.
2. **Goding, J. W.,** *Monoclonal Antibodies: Principles and Practice,* Academic Press, New York, 1983, 59.
3. **Slavin, D.,** Production of antisera in rabbits using calcium alginate as an antigen depot, *Nature (London),* 165(4186), 115, 1950.
4. **van Preis, I. and Langer, R.,** A single-step immunization by sustained antigen release, *J. Immunol. Methods,* 28, 193, 1979.
5. **Goding, J. W.,** *Monoclonal Antibodies: Principles and Practice,* Academic Press, New York, 1983, 57.
6. **Goding, J. W.,** *Monoclonal Antibodies: Principles and Practice,* Academic Press, New York, 1983, 58.
7. **Goding, J. W.,** *Monoclonal Antibodies: Principles and Practice,* Academic Press, New York, 1983, 60.
8. **Goding J. W.,** *Monoclonal Antibodies: Principles and Practice,* Academic Press, New York, 1983, 62.
9. **Goding, J. W.,** *Monoclonal Antibodies: Principles and Practice,* Academic Press, New York, 1983, 64.
10. **Galfre, G. and Milstein, C.,** Preparation of monoclonal antibodies: strategies and procedures, *Methods Enzymol.,* 73, 1, 1981.
11. **Goding, J. W.,** *Monoclonal Antibodies: Principles and Practice,* Academic Press, New York, 1983, 67.
12. **Goding, J. W.,** *Monoclonal Antibodies: Principles and Practice,* Academic Press, New York, 1983, 73.
13. **Goding, J. W.,** *Monoclonal Antibodies: Principles and Practice,* Academic Press, New York, 1983, 75.
14. **Kekwick, R. A.,** The serum proteins in multiple myelomatosis, *Biochem. J.,* 34, 1248, 1940.
15. **Peterson, E. A.** Myelomatosis, Chromatography of proteins. I. Cellulose ion-exchange adsorbent, *J. Am. Chem. Soc.,* 78, 751, 1956.
16. **Parham, P.,** Monoclonal antibodies against HLA products and their use in immunoaffinity purification, *Methods Enzymol.,* x92E, 110, 1983.
17. **Bouvet, J. P., Pires, R., and Pillot, J.,** A modified gel filtration technique producing an unusual exclusion volume of IgM: a simple way of preparing monoclonal antibodies, *J. Immunol. Methods,* 66 (2), 299, 1984.
18. **Jeppson, J. O., Lawell, C. B., and Franzen, B.,** Agarose gel electrophoresis, *Clin. Chem.,* 25, 629, 1979.
19. **Kolin, A.,** Separation and concentration of proteins in a pH field combined with an electric field, *J. Chem. Phys.,* 22, 1628, 1954.
20. **Campbell, D. H., Leuscher, E., and Lerman, L. S.,** Immunological adsorbents. I. Isolation of antibody by means of a cellulose protein antigen, *Proc. Natl. Acad. Sci. U.S.A.,* 37, 575, 1951.
21. **Bjorn, M. J., Ring, D., and Frankel, A.,** Evaluation of monoclonal antibodies for the development of breast cancer immunotoxins, *Cancer Res.,* 45, 1214, 1985.
22. **Cushley, W.,** Immunotoxins — immunochemical engineering for novel therapeutic reagents, *Pharma Int.,* p. 33, February 1985.
23. **Kulkarni, P. N., Blair, A. H., and Ghose, T. I.,** Covalent binding of methotrexate to immunoglobulins and the effect of antibody-linked drug on tumor growth *in vivo, Cancer Res.,* 41, 2700, 1981.
24. **Arnold, L. J., Jr., Dugan, A., and Kaplan, N. O.,** Poly (L-lysine) as an antineoplastic agent and tumor-specific drug carrier, in *Targeted Drugs,* Goldberg, E. P., Ed., Interscience, New York, 1983, 89.
25. **Arnon, R. and Hurwitz, E.,** Antibody- and polymer-drug conjugates, in *Targeted Drugs,* Goldberg, E. P., Ed., Interscience, New York, 1983, 23.
26. **Manabe, Y., Tsubota, T., Haruta, Y., Kataoka, K., Okazaki, M., Haisa, S., Nakamura, K., and Kimura, I.,** Production of a monoclonal antibody-methotrexate conjugate utilizing dextran T-40 and its biologic activity, *J. Lab. Clin. Med.,* 104(3), 445, 1984.
27. **Suzuki, T., Sato, E., Goto, Y., Katsurada, Y., Unno, K., and Takahashi, T.,** The preparation of mitomycin C, andriamycin and daunomycin covalently bound to antibodies as improved cancer chemotherapeutic agents, *Chem. Pharm. Bull.,* 29(3), 844, 1981.
28. **Hurwitz, E., Levy, R., Maron, R., Wilchek, M., Arnon, R., and Sela, M.,** The covalent binding of daunomycin and adriamycin to antibodies, with retention of both drug and antibody activities, *Cancer Res.,* 35, 175, 1975.
29. **Gallego, J., Price, M. R., and Baldwin, R. W.,** Preparation of four daunomycin-monoclonal antibody 791T/36 conjugates with anti-tumor activity, *Int. J. Cancer,* 33, 737, 1984.
30. **Varga, J. M., Asato, N., Lande, S., and Lerner, A. B.,** Melanotropin-daunomycin conjugate shows receptor-mediated cytotoxicity in cultured murine melanoma cells, *Nature (London),* 267, 56, 1977.
31. **Rowland, G. F.,** The use of antibodies in drug targeting and synergy, in *Targeted Drugs,* Goldberg, E. P., Ed., Interscience, New York, 1983, 57.
32. **Manabe, Y., Tsubota, T., Haruta, Y., Kataoka, K., Okazaki, M., Haisa, S., Nakamura, K., and Kimura, I.,** Production of a monoclonal antibody-mitomycin C conjugate utilizing dextran T-40 and its biologic activity, *Biochem. Pharmacol.,* 34(2), 289, 1985.

33. **Hwang, K. M., Foon, K. A., Cheung, P. H., Pearson, J. W., and Oldham, R. K.,** Selective antitumour effect on L10 hepatocarcinoma cells of a potent immunoconjugate composed of the A chain of abrin and a monoclonal antibody to a hepatoma-associated antigen, *Cancer Res.,* 44, 4578, 1984.

34. **Youle, R. J. and Neville, D. M., Jr.,** Anti-Thy 1.2 monoclonal antibody linked to ricin is a potent cell-type specific toxin, *Proc. Natl. Acad. Sci. U.S.A.,* 77(9), 5483, 1980.

35. **Ukun, F. M., Ramakrishnan, S., and Houston, L. L.,** Increased efficiency in selective elimination of leukemia cells by a combination of cyclophosphamide and a human B-cell-specific immunotoxin containing pokeweed antiviral protein, *Cancer Res.,* 45, 69, 1985.

36. **Weil-Hillman, G., Runge, W., Jansen, F. K., and Vallera, D. A.,** Cytotoxic effect of anti-M_r 67,000 protein immunotoxins on human tumors in a nude mouse model, *Cancer Res.,* 45, 1328, 1985.

37. **McIntosh, D. P., Edwards, D. C., Cumber, A. J., Parnell, G. D., Dean, C. J., Ross, W. C. J., and Forrester, J. A.,** Ricin B chain converts a non-cytotoxic antibody ricin A chain conjugate into a potent and specific cytotoxic agent, *FEBS Lett.,* 164(1), 17, 1983.

38. **Cushley, W.,** Immunotoxins — immunochemical engineering for novel therapeutic reagents, *Pharma Int.,* p. 36, February 1985.

39. **Kishida, K., Masuho, Y., Saito, M., Hara, T., and Fuji, H.,** Ricin A-chain conjugated with monoclonal anti-L1210 antibody *in vitro* and *in vivo, Cancer Immunol. Immunother.,* 16, 93, 1983.

40. **O'Neil, G. J.,** The use of antibodies as drug carriers, in *Drug Carriers in Biology and Medicine,* Gregoriadis, G., Ed., Academic Press, London, 1979. chap.2.

41. **Manabe, Y., Tsubota, T., Haruta, Y., Kataoka, K., Okazaki, M., Haisa, S., Makamura, K., and Kimura, I.,** Production of monoclonal antibody-methotrexate conjugate utilizing dextran T-40 and its biologic activity, *J. Lab. Clin. Med.,* 104(3), 450, 1984.

42. **Ramakrishnan, S. and Houston, L. L.,** Comparison of selective cytotoxic effects of immunotoxins containing ricin A chain or pokeweed antiviral protein and anti-Thy 1.1 monoclonal antibodies, *Cancer Res.,* 44, 201, 1984.

43. **Mizusawa, E., Dahlman, H. L., Bennett, S. J., Goldberg, D. M., and Hawthrone, M. F.,** Neutron-capture therapy of human cancer: *in vitro* results on the preparation of boron-labeled antibodies to carcinoembryonic antigen, *Proc. Natl. Acad. Sci. U.S.A.,* 79, 3011, 1982.

44. **Miescher-Granger, S., HochKeppel, H. K., Braun, D. G., and Alkan, S. S.,** Biological activities of human recombinant interferon a/B targeted by anti-Epstein-Barr virus monoclonal antibodies, *FEBS Lett.,* 179(1,) 29, 1979.

45. **Hurwitz, E., Arnon, R., Sahar, E., and Danon, Y.,** A conjugate of adriamycin and monoclonal antibodies to Thy-1 antigen inhibits human neuroblastoma cells *in vitro, Ann. N.Y. Acad. Sci.,* 417, 125, 1983.

46. **Thorpe, P. E. and Ross, W. C. J.,** The preparation and cytotoxic properties of antibody-toxin conjugates, *Immunol. Rev.,* 62, 119, 1982.

47. **Bernstein, I. D., Tam, M. R., and Nowinski, R. C.,** Mouse leukemia: therapy with monoclonal antibodies against a thymus differentiation antigen, *Science,* 207, 68, 1980.

48. **Kramer, P. A.,** Albumin microspheres as vehicles for achieving specificity in drug delivery, *J. Pharm. Sci.,* (63)10, 1646, 1974.

49. **Sugibayashi, K., Morimoto, Y., Nadai, T., and Kato, Y.,** Drug-carrier property of albumin microspheres in chemotherapy. I. Tissue distribution of microsphere-entrapped 5-fluorouracil in mice, *Chem. Pharm. Bull.,* (25)12, 3433, 1977.

50. **Longo, W. E., Iwata, H., Lindheimer, T. A., and Goldberg, E. P.,** Preparation of hydrophilic albumin microspheres using polymeric dispersing agents, *J. Pharm. Sci.,* (71)12, 1323, 1982.

51. **Sugibayashi, K., Morimoto, Y., Nadai, T., Kato, Y., Hasegawa, A., and Ariya, T.,** Drug-carrier property of albumin microspheres in chemotherapy. II. Preparation and tissue distribution in mice of microsphere-entrapped 5-fluorouracil, *Chem. Pharm. Bull.,* (27)1, 204, 1979.

52. **Martodam, R. R., Ywumasi, D. Y., Liener, L. E., Powers, J. C., Nishino, N., and Krejcarek, C.,** Albumin microspheres as carriers of an inhibitor of leukocyte elastage: potential therapeutic agent for emphysema, *Proc. Natl. Acad. Sci. U.S.A.,* (76)5, 2128, 1979.

53. **Goosen, F. A., Leung, Y. F., O'Shea, G. M., Chou, S., and Sun, A. M.,** Long acting insulin: Slow release of insulin from a biodegradable matrix implanted in diabetic rats, *Diabetes,* 32, 748, 1983.

54. **Senyei, A., Widder, K., and Czerlinski, G.,** Magnetic Guidance of drug-carrying microspheres, *J. Appl. Phys.,* (49)6, 3578, 1978.

55. **Tuma, R. F., Forsberg, J. O., and Agerup, B.,** Enhanced uptake of actinomycin D in dog kidney by simultaneous injection of degradable starch microspheres in the renal artery, *Cancer,* 50(1), 1, 1982.

56. **Lindell, B., Aronsen, K. F., Nosslin, B., and Rothman, U.,** Studies in pharmacokinetics and tolerance of substances temporarily retained in the liver by mircosphere embolization, *Ann. Surg.,* 187(1), 95, 1978.

57. **Forsberg, J. O.,** Transient blood flow reduction induced by inter-arterial injection of degradable starch microspheres, *Acta Chir. Scand.,* 144, 275, 1978.

58. **Tuma, R. F., Forsberg, J. O., Schosser, R., and Arfors, K. E.,** The trapping of drugs in microcirculation with degradable microspheres, *Bibl. Anat.,* 18, 210, 1978.

59. **Arfors, K. E., Forsberg, J. O., Larsson, B., Lewis, D. H., Rosengren B., and Odman, S.,** Temporary intestinal hypoxia induced by degradable microspheres, *Nature (London),* 262, 500, 1976.

60. **Wright K., Anderson, J. H., Giantures, C., Wallace, S., and Chuang, V. P.,** Partial splenic embolization using polyvinyl alcohol foam, dextran, polystryene, or silicone, *Radiology,* 142, 351, 1982.

61. **Widder, K., Flouret, G., and Senyei, A.,** Magnetic microspheres: synthesis of a novel parenteral drug carrier, *J. Pharm. Sci.,* 68(1), 79, 1979.

62. **Morimoto, Y., Sugibayashi, K., and Akimoto, M.,** Magnetic guidance of ferro-colloid-entrapped emulsion for site-specific drug delivery, *Chem. Pharm. Bull.,* 31(1), 279, 1983.

63. **Gregoriadis, G. and Neerunjun, D. E.,** Control of the rate of hepatic uptake and catabolism of liposome-entrapped proteins injected into rats. Possible therapeutic applications, *Eur. J. Biochem.,* 47, 179, 1974.

64. **Gregoriadis, G.,** The carrier potential of liposomes in biology and medicine. *N. Engl. J. Med.,* 295(13), 704, 2976.

65. **Wu, P., Wu, H., Tin, G. W., Schuh, J. R., Croasmun, W. R., Baldeschwieler, J. D., Shan, T. Y., and Ponpipom, M. M.,** Stability of carbohydrate-modified vesicles *in vivo:* comparative effects of ceramide and cholesterol glycoconjugates, *Proc. Natl. Acad. Sci. U. S. A.,* 79, 5490, 1982.

66. **Juliano, R. L. and Stamp, D.,** The effect of particle size and charge on the clearance rates of liposomes and liposome encapsulated drugs, *Biochem. Biophys. Res. Commun.,* 63(3), 651, 1975.

67. **Yatvin, M. B., Weinstein, J. N., Dennis W. H., and Blumenthal, R.,** Design of liposomes for enhanced local release of drugs by hyperthermia, *Science,* 202(22), 1290, 1978.

68. **Yatvin, M. B., Kreutz, W., Horwitz, B. A., and Shinitzky, M.,** pH-sensitive liposomes: possible clinical applications, *Science,* 210(12), 1253, 1980.

69. **Rawls, R.,** New methods let drugs past blood-brain barrier, *Chem. Eng. News,* p. 24, 1981.

70. **Bodor, N., Farag, H. H., Brewster, M. E., III,** Site specific, sustained release of drugs to the brain, *Science,* 214(18), 1370, 1981.

71. **Bodor, N. and Farag, H. H.,** Improved delivery through biological membranes. XIV. Brain-specific, sustained delivery of testosterone using a redox chemical delivery system, *J. Pharm. Sci.,* 73(3), 385, 1984.

72. **Ringsdorf, H.,** in *Polymeric Delivery Systems,* Kostelnick, R. J., Ed., Gordon and Breach, New York, 1978, 197.

73. **Trouet, A.,** in *Polymeric Delivery Systems,* Kostelnick, R. J., Ed., Gordon and Breach, New York, 1978, 157.

74. **Samour, C. M.,** in *Polymeric Drugs,* Donaruma, L. G. and Vogl, O., Eds., Academic Press, New York, 1978, 161.

75. **Goldberg, E. P.,** in *Polymeric Delivery Ssytems,* Kostelnick, R. J., Ed., Gordon and Breach, New York, 1978, 227.

76. **Goldberg, E. P.,** in *Polymeric Drugs,* Donaruma, L. G. and Vogl, O., Academic Press, New York, 1978, 239.

77. **Rowland, G. F., O'Neill, G. J., and Davies, D. A. L.,** Supression of tumour growth in mice by a drug-antibody conjugate using a novel approach to linkage, *Nature (London),* 255, 487, 1975.

78. **Chu, B. F. and Whiteley, J. M.,** High molecular weight derivatives of methotrexate as chemotherapeutic agents, *Mol. Pharmacol.,* 13, 80, 1977.

79. **Tyrrell, D. A. and Ryman, B. E.,** The entrapment of therapeutic agents in resealed erythrocyte "ghosts" and their fate *in vivo, Biochem. Soc. Trans.,* 4, 677, 1976.

80. **Wilson, W. D. and Jones, R. L.,** Intercalating drugs: DNA binding and molecular pharmacology, in *Advances in Pharmacology and Chemotherapy,* Vol. 18, Schnitzer, R. J., Ed., Academic Press, New York, 1981, 177.

81. **Sjoholm, I. and Edman, P.,** Acrylic microspheres *in vivo.* I. Distribution and elimination of polyacrylamide microparticles after intravenous and intraperitoneal injection in mouse and rat, *J. Pharmacol. Exp. Ther.* 211(3), 656, 1979.

82. **Edman, P., Sjoholm, I., and Brunk, U.,** Ultrastructural alterations in macrophages after phagocytosis of acrylic microspheres, *J. Pharm. Sci.,,* 73(2), 153, 1984.

83. **Edman, P., Sjoholm, I., and Brunk, U.,** Acrylic microspheres *in vivo.* VII. Morphological studies on mice and cultured macrophages, *J. Pharm. Sci.,* 72(6), 658, 1983.

84. **Stendahl, O., Dahlgren, C., Edebo, M., and Ohman, L.,** in *Endocytosis and Exocytosis in Host Defense,* Vol. 17, S. Karger, New York, 1981, 12.

85. **Arturson, P., Laakso, T., and Edman, P.,** Acrylic microspheres *in vivo.* XI. Blood elimination kinetics and organ distribution of microparticles with different surface characteristics, *J. Pharm. Sci.,* 72(12), 1415, 1983.

86. **Zaffaroni, A. and Bonsen, P.,** in *Polymeric Drugs,* Donaruma, L. G. and Vogl, O., Eds., Academic Press, New York, 1978, 1.

87. **Kalal, J., Drobnik, J., Kopecek, J., and Exner, J.,** in *Polymeric Drugs,* Donaruma, L. G. and Vogl, O., Eds., Academic Press, New York, 1978, 131.

Chapter 9

STERILIZATION OF CONTROLLED RELEASE SYSTEMS

Thomas H. Ferguson

TABLE OF CONTENTS

I. INTRODUCTION

Sterilization may be defined as a process that achieves total destruction or removal of all living microorganisms, including bacterial and fungal spores. In contrast, disinfection may be defined as any process that destroys pathogens in the nonsporulating or vegetative state. The object of sterilization is to assure that pathogens do not enter the body when material is administered parenterally, sterilization being essential whenever the skin integrity is breached.

Except for the sterilization process of microfiltration, normal sterilization methods involve killing microorganisms rather than removing them. Populations of microorganisms are killed exponentially when exposed to common sterilization methods. Therefore, increased time cannot achieve theoretical absolute sterility. The number of bacterial cells dying is a function of the number of survivors present. The number of survivors is dependent upon the initial number of microorganisms present. Thus, prior to any sterilizaton process, the initial number of microorganisms should be kept to a minimum.

The exponential death process can be described by the following relationship:

$$N_t = N_o e^{-kt} \tag{1}$$

where t = time, N_o = initial number of microorganisms, N_t = number of live microorganisms remaining at time t, and k = rate of death constant. The rate of death can be expressed as:

$$k = 1/t \ln N_o/N_t \tag{2}$$

and usually results in a straight line when $\ln N_o/N_t$ is plotted vs. time. However, in many cases, the rate of death of microorganisms is not linear, but sigmoid or of some other shape. This may be due to the initial delayed reaction of the sterilization agent with the microbial protoplasm, while the decrease in the rate of death at the end of the sterilization period may be due to the development or survival of resistant mutants.[1]

There are alternative ways of expressing the death rate of microorganisms, such as the decimal reduction time (D) and the lethality factor (F). The D value is the exposure time under a particular sterilization condition to reduce the microbial population by 90% or one logarithm and is the reciprocal of the death rate constant, k. The F value, normally used in conjunction with steam sterilization, is the time (usually in minutes) at a particular temperature to achieve an equivalent sterilization effect (probability of survival) to that provided at 121°C.

No single sterilization method has been developed for universal use due to the different effects that a sterilization process has on different materials. However, numerous sterilization methods have been available and utilized for some time for medical devices, pharmaceuticals, and food processing. As the rapidly emerging field of controlled drug release develops, the achievement of sterility by the proper method and its effects on the controlled release system will become vital. Factors such as the physical appearance, color, crystalline form, particle size, solubility, pH, or palatability of the drug can be affected by the sterilization method. The preservation of drug potency is tacit. More importantly, the effects of sterilization on the materials or polymers used to achieve controlled release of the drug, such as the physical appearance, color, mechanical properties, and the effects on the rate-controlling parameters governing controlled release should be evaluated. As an increasing number of novel controlled release systems become available and attempts are made to commercialize these systems, the sterilization method (if required) will become an integral part in the development process. Thus, the sterilization method should be addressed in the primary stages of development. The purpose of this chapter is to outline the various sterilization methods currently available, review the effects of sterilizing processes on selected drugs and materials, and discuss the relevance of these effects to controlled release systems.

II. STERILIZATION METHODS

In general, sterilization methods can be classified into two categories: physical methods and chemical methods. The physical methods include thermal and nonthermal processes, whereas the chemical methods include gaseous and liquid processes. The physical methods are usually more economical than the chemical methods.[2]

A. Physical Methods
1. Thermal
a. Dry Heat
Hot air or dry heat kills microbes by oxidation. This sterilization method utilizes a conventional fan-equipped, hot-air oven. Rapid throughput can be achieved by using conveyor ovens. Recommended sterilization conditions range from 1.5 min at 190°C to 45 min at 160°C.[2] USP recommends not less than 2 hr at 160 to 170°C.[3] Higher temperatures and corresponding shorter times may be used for heat-resistant materials, while lower temperatures and longer periods may be best for heat-sensitive materials. The exact times and temperatures are dependent on the nature of the product to be sterilized, the size of the individual containers, and their distribution in the oven. Maintenance of a uniform temperature distribution is critical. Since the heat transfer in dry heat sterilization is slow, the economic advantages of a high-temperature short-duration exposure may not be realized as not all articles in a load may be maintained at the essential temperature and time period to achieve sterilization.

b. Moist Heat
Moist heat or steam heat kills microbes by denaturing their proteins.[4] Saturated steam under pressure is probably the most widely used general sterilization process in hospitals today because of effectiveness, simplicity, and economics. Efficient steam heat depends on the pressure and on the latent heat of vaporization given up on condensation. Because steam is able to give up its latent heat without a drop in temperature, it is superior to dry heat sterilization at a given temperature. Recommended sterilizing conditions of temperature and time are 3 min at 134°C (30 psi), 10 min at 126°C (20 psi), and 15 min at 121°C (15 psi).[4] USP recommends not less than 15 min at 121°C.[3] These conditions are for steam exposure and do not include the time necessary to penetrate the load. Steam heat penetrates well through materials such as fabrics, paper, and some plastic films, but is not applicable on sealed, impermeable containers unless these contain moisture, e.g., glass-sealed aqueous solutions. In addition, oily liquids and powders cannot be sterilized by steam heat. For these cases, steam heat is inferior to dry heat or radiation for sterilization.[2]

2. Nonthermal
a. UV Radiation
Historically, UV radiation has not been used as it only achieves surface sterilization. In some cases it was only bacteriostatic, not killing vegetative and spore forms of microorganisms. This was probably due to UV doses of insufficient magnitude since recent reports utilizing high-intensity UV radiation have shown it to be effective in killing both vegetative and spore forms of microorganisms (see Table 1). The maximum bactericidal effect occurs at wavelengths between 240 and 280 nm, with the peak efficiency at 253.7 nm.[5] The lethal action of UV radiation is due to the formation of pyrimidine dimers from adjacent monomers on the same deoxyribonucleic acid (DNA) strand and indirectly due to the production of

Table 1
BACILLUS PUMILUS SPORE SURVIVOR
DATA FOR UV RADIATION[5]

Exposure time (sec)	UV dose[a] (erg/mm^2)	Log$_{10}$ spore count		
		5 cc	15 cc	30cc
0	0	5.8921	5.9494	5.8261
5	137,000	5.6021	5.1761	4.8092
10	275,000	5.5315	4.2788	3.9174
15	412,500	4.0792	3.0792	3.0289
20	550,000	2.5682	2.5185	2.1404
25	687,000	0.0000	0.0000	0.0000
30	825,000	0.0000	0.0000	0.0000

[a]   UV dose delivered to duplicate sets of spore suspensions prepared in 5-, 15-, and 30-cc low density polyethylene (LDPE) bottles. UV density and wavelength at target position were 275,000 μW/cm^2 and 253.7 nm, respectively, for the 5- and 15-cc LDPE bottles. For the 30-cc LDPE bottles, the UV density at target position was 250,000 μW/cm^2.

Table 2
GAMMA STERILIZATION VS. ELECTRON BEAM STERILIZATION[8,9]

Variable	^{60}Co	^{137}Cs	Electron beam
Source	Gamma rays	Gamma rays	Electrons
Penetration	High, 1.1 and 1.3 MeV/ disintegration	High, 0.67 MeV/disintegration	Moderate
Dose rate	<1 Mrad/hr	<1 Mrad/hr	1 Mrad/sec
Product dwell	~8 hr	>8 hr	Minutes
Throughput	~1 case/min	<1 case/min	~7 cases/min
Case size	Fixed	Fixed	Flexible
Installation	Permanent	Permanent	Movable
Product temp	5°C rise/Mrad	5°C rise/Mrad	Room temp
Source variation	Decays 1%/month, $t_{1/2}$ = 5.3 years	Decays < 1%/month, $t_{1/2}$ = 33 years	None

peroxides in the medium which act as oxidizing agents.[6] Because UV radiation has only two control parameters, intensity and exposure time, and is economical, it appears that high-intensity UV radiation may find increased use as a specialized sterilization method in the future.

b. Ionizing Radiation

Ionizing radiation is one of the simplest and most effective means of sterilization. For sterilization purposes the ionizing radiations of practical importance are high-energy electrons from high-voltage accelerators and gamma rays from ^{60}Co or ^{137}Cs sources. These methods are efficient and clean. Greater dose rates are achievable with electron beams; consequently, shorter exposure times are required than with gamma rays. However, gamma rays have more penetration power than electron beams. In contrast to gamma radiation, high-energy electrons with energies greater than 5 MeV induce radioactivity in materials.[1] Even with the higher installation costs of gamma radiation, gamma ray sterilization has found acceptance as a sterilization method due to its superior penetration abilities. The characteristics for ^{60}Co, ^{137}Cs, and electron beam radiation are summarized in Table 2.

Table 3
RESISTANCE OF MICROORGANISMS TO ⁶⁰Co
RADIATION[10]

Microorganism	D value (Mrad)	MID (Mrad)
USP pathogens		
Escherichia coli	0.020	0.12
Salmonella		
typhimurium	0.020	0.12
senftenberg	0.013	0.08
Staphylococcus aureus	0.016	0.10
Pseudomonas aeruginosa	0.013	0.08
Other vegetative cells		
Diplococcus pneumoniae	0.013	0.08
Hemophilus influenzae	0.040	0.24
Proteus vulgaris	0.071	0.43
Sarcina lutea	0.045	0.27
Streptococcus fecalis	0.050	0.30
Spore formers		
Bacillus pumilus	0.192	1.15
B. subtilis		
Spores	0.190	1.14
Vegetative cells	0.025	0.15
Clostridium botulinum		
Type A	0.350	2.10
Type B	0.200	1.20
Clostridium tetani	0.150	0.90
C. sporogenes	0.210	1.26
Fungi and molds		
Aspergillus niger	0.045	0.27
Penicillus notatum	0.085	0.51

The accepted sterilization dose is 2.5 Mrad, the rad corresponding to the absorption of 100 erg/g of material. However, many radiation-sensitive articles having low or susceptible bioburdens can be effectively sterilized at lower dose levels. D values and the minimum irradiation dose (MID) required to achieve 10^{-6} probability of nonsterility for various microorganisms are shown in Table 3.

Bruck[1] states that two theories have been proposed for the lethal action of ionizing radiation on microorganisms. The direct action theory postulates that the radiation induces ionization of the DNA of the microorganism. The indirect action theory suggests that the primary step involves the formation of free radicals, e.g., peroxides, or other molecules in the medium, subsequently inducing secondary reactions in the DNA of the microorganism.[1,6,7]

c. Microfiltration

Microfiltration is a specialized technique for low-viscosity liquids susceptible to heat or other sterilization methods and depends upon the physical removal of microorganisms, but not their soluble metabolic products, by adsorption onto a filter or sieve mechanism. The separation of microorganisms from the filtrate may involve interactions associated with electrostatic forces or mechanical sieving by the size, shape, and tortuousness of the voids. Suitable filters achieving sterility have a nominal porosity of 0.22 μm. With the appropriate choice of smaller porosity filters, even viruses can be excluded.

B. Chemical Methods
1. Gases

Numerous gases have been investigated as sterilizing agents: ethylene oxide, ozone,

Table 4

CHARACTERISTICS OF GASEOUS STERILIZATION AGENTS[1,2,4,7,11,12]

Characteristic	Ethylene oxide (ETO)	Propylene oxide	Ozone	Formaldehyde
Boiling point, °C	10.7	34.2	−111.9	−19.5
Freezing point, °C	−111	−112	−193	−92
Penetration	Very rapid	<ETO	Rapid	Limited, humidity needed
Effectiveness	100%	Half of ETO	Dependent on temperature and humidity	Dependent on humidity, tends to polymerize on surface
Danger	Explosive	Flammable	Explosive	Flammable
Hazard	Highly irritating to eyes, mucous membranes, carcinogen	$LD_{50} = 1.14$ g/kg	Irritation of respiratory tract, eyes	Irritating to mucous membranes, carcinogen

propylene oxide, formaldehyde, β-propiolactone, ethyleneimine, peracetic acid, methyl bromide, chloropicrin dioxide, and chlorine.[1,2] Characteristics of four major gaseous sterilization agents are summarized in Table 4. In practice, gas sterilization most commonly uses ethylene oxide mixed with dichlorofluoromethane (Freon 12) or carbon dioxide to reduce the explosiveness of ethylene oxide in air. It has been shown that mixtures of Freon 12 and ethylene oxide are nonflammable in any proportion of air at temperatures up to 54.4°C.[1]

Ethylene oxide is active against all microorganisms including viruses and bacterial spores.[2] Ethylene oxide is a very potent alkylating agent and probably destroys microorganisms by alkylation of the susceptible sulfhydryl groups in proteins:[11]

$$CH_2OCH_2 + \text{protein} - SH \rightarrow \text{protein}-S-CH_2-CH_2OH$$

The resultant interference with cell metabolism is irreversible and bactericidal. Ethylene oxide can react with other functional groups as follows:

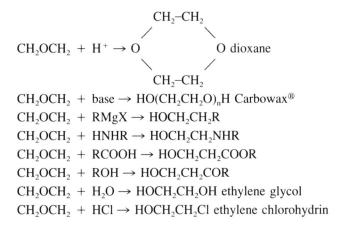

$$CH_2OCH_2 + \text{base} \rightarrow HO(CH_2CH_2O)_nH \text{ Carbowax}®$$
$$CH_2OCH_2 + RMgX \rightarrow HOCH_2CH_2R$$
$$CH_2OCH_2 + HNHR \rightarrow HOCH_2CH_2NHR$$
$$CH_2OCH_2 + RCOOH \rightarrow HOCH_2CH_2COOR$$
$$CH_2OCH_2 + ROH \rightarrow HOCH_2CH_2COR$$
$$CH_2OCH_2 + H_2O \rightarrow HOCH_2CH_2OH \text{ ethylene glycol}$$
$$CH_2OCH_2 + HCl \rightarrow HOCH_2CH_2Cl \text{ ethylene chlorohydrin}$$

The reaction that is bactericidal in microorganisms also takes place with human protein. Thus, appropriate care must be taken when using ethylene oxide to protect the operator and to completely remove residual amounts of ethylene oxide in the sterilized articles. The two major by-products of ethylene oxide sterilization, ethylene glycol and ethylene chlorohydrin, are also absorbed into the sterilized article and must be removed. Both ethylene glycol and ethylene chlorohydrin can form *in situ* in the body if residual ethylene oxide is present.

Thus, when sterilizing by ethylene oxide proper aeration (a diffusion-controlled process) of the sterilized article is essential. Unless the maximum tolerable levels of ethylene oxide and its by-products are defined, complete elimination must be attempted.

2. *Liquids*

Treatment by liquid chemicals is not a very successful general method of sterilization except in special applications. In most cases, the sterilizing agent used is dissolved into an aqueous solvent. When the article to be sterilized is immersed into the aqueous sterilizing solution, small amounts of the solution will be absorbed dependent upon the hydrophilicity of the item.[13] For hydrophilic articles, the sterilizing solution permeates the item and it is difficult to remove all of the sorbed sterilizing agent under sterile conditions. For hydrophobic articles, the aqueous sterilizing solution may not wet the surface properly and inadequately penetrates small cavities or narrow pores, resulting in incomplete sterilization. Because of these practical problems, liquid chemicals are thought of as disinfectants rather than true sterilants. Many liquid sterilants (disinfectants) have been utilized, including mercurials, phenols, quaternary ammonia compounds (Zephiran), chlorine compounds, iodine, formaldehyde solutions, alcohols, glutaraldehyde, hexachlorophene, and hydrogen peroxide.

C. Summary

It is interesting to note that of the numerous sterilization methods available today, ethylene oxide sterilization by far is the most widely used. Of 1260 Good Manufacturing Practices (GMP) inspections performed by the Food and Drug Administration (FDA) in 1980, which included 200 sterile device manufacturers, 64% used ethylene oxide, 20% used steam, 8% used radiation, and 8% used other sterilization methods such as sterile filtration and liquid chemicals.[14] Since there are several continuous radiation sterilization installations with high throughput rates, the actual number of articles sterilized (compared to the number of firms using radiation) is most likely greater than 8%. Concern for the carcinogenic potential of ethylene oxide, sterilization efficacy, and the problems in achieving low ethylene oxide and by-product residues in the sterilized article is causing the U.S. regulatory agencies (FDA and Environmental Protection Agency)[15,16] and the medical device and pharmaceutical industries to review the use of ethylene oxide sterilization. The result of this review will probably mean the increased usage of alternative sterilization methods, in particular radiation sterilization. It is axiomatic that the most simple process and ease of control reduces the number of things that can go wrong, resulting in improved efficiency and reproducibility; this is the advantage of radiation sterilization (see Table 5).

III. STERILIZATION EFFECTS ON DRUGS

The release kinetics from a controlled release system, as well as drug bioavailability, is a function of the properties of the drug. Sterilization-induced changes in physicochemical properties such as the dissolution rate, particle size, solubility, crystalline form, pH, diffusion coefficient, and partition coefficient can significantly modify drug performance and may even prohibit incorporation of the drug into a controlled release system. Other factors such as color changes, and in the case of oral dosage forms, changes in palatability, although not directly affecting bioavailability and release kinetics, indirectly indicate that one or more of the physicochemical properties may have changed. It is essential that in establishing a sterilization method the effects on the physicochemical properties of the drug be evaluated. Tacit to this discussion is the maintenance of the efficacy and toxicity profiles of the drug.

The amout of water present with the drug in a controlled release system is critical. With ethylene oxide sterilization, ethylene glycol can be formed, possibly modifying the dissolution rate, solubility, diffusion coefficient, and crystalline form of the drug when the ethylene

Controlled Release Systems: Fabrication Technology

Table 5
CONTROLS AFFECTING STERILIZATION METHODS[5,8,17]

Control	Dry heat	Steam	UV radiation	Ionizing radiation	Micro-filtration	Ethylene oxide	Liquids
Temperature	*	*				*	*
Time	*	*	*	*	*	*	*
Pressure		*			*	*	
Vacuum		*			*	*	
Humidity		*				*	
Concentration					Area	*	*
pH							*
Packaging	*	*		a	Asepsis	*	*
Poststerilization drying/degassing		*				*	
Residual toxicity		b	b	b		b	
Type of process	Batch	Batch	Batch, possible continuous	Continuous	Batch	Batch	Batch

Note: a, In the final impermeable packaging; b, depends on the type of materials sterilized.

glycol is completely removed. With regard to radiation sterilization, water acts as an energy transfer agent. The primary radiolysis products of water, the hydrated electron, e_{aq}^-, $\cdot$OH, and H$\cdot$, may interact and transfer its acquired energy to a solute molecule. In dilute solutions, it is thought that most of the energy is imparted into the water and, therefore, most of the radiation damage occurs via transfer of the radiolysis products.[18] None of the radiolysis products are stable in pure water; the resultant final products are water, hydrogen peroxide, and hydrogen gas.

The study of benzylpenicillin in an aqueous solution and in a powder illustrates the difference in radiation response between aqueous and dry preparations. Blackburn et al.[19] reported extensive degradation of benzylpenicillin when irradiated in dilute (10^{-4} to 10^{-2} M) aqueous solutions. They suggested that the extensive degradation was caused by two of the radiolysis products of water, e_{aq}^- and $\cdot$OH. The hydroxyl radical was implicated in the hydroxylation of the benzene ring in the penicillin side chain, the formation of benzylpenicilloic acid and benzylpenillic acid, and reaction with the amide side chain to produce phenylacetic acid and the o-, m-, p-hydroxybenzylpenicillins. The hydrated electron led to formation of benzylpenillic acid and benzylpenilloic acid. In contrast, the radiation stability of dry benzylpenicillin was extremely good with negligible decomposition below a 3-Mrad dose. Other penicillins are susceptible to radiolytic decarboxylation in the dry state.[18,19] Blackburn et al.[19] showed that five typical sulfonamides were extensively degraded in aqueous solution, whereas in the dry state they were resistant to decomposition to a sterilizing dose of 2.5 Mrad.

Provided that the degradation pathways of a drug are known in radiation sterilization, protection from degradation can be achieved by the addition of radical scavengers, assuming that the scavengers do not affect the efficacy of the drug. This can be achieved by using $-$SH-containing molecules, scavengers of radiolysis products of water, or reagents that convert radiolysis products to the parent compound. From a kinetic competition viewpoint, the protective effect of the radical scavengers will be the greatest when the drug is present in low concentrations. Kaetsu et al.[20] suggest that the degradation of chemotherapeutic drugs by radiation can be retarded by irradiating at low temperatures ($< -40°$C) and at a dose of less than 0.5 Mrad. However, this dose may not be sufficient to achieve sterility. Kaetsu et al.[20] and Tsuji et al.[10] also suggest that the degradation of drugs when exposed to radiation

Table 6
RADIATION STERILIZATION EFFECTS ON DRUGS

Compounds	Preparation[a]	Effect	Ref.
Steroids	0	Stable; degrades	10,18
	1	Degrades	18
	4	Stable	10
Carbohydrates	0	Degrades	18
	1	Degrades	18
Tetracyclines	0	Minimal degradation	18
	2	Stable	17
Penicillins	0	Stable to minimal degradation	18, 19
Cephalosporins	0	Stable to minimal degradation	18
Vitamin B_{12}	1	Degrades	19
Indole	1	Degrades	19
Fluorescein sodium	1	Stable	17
	3	Stable	17
Urea	1	Stable	17
Ethylmorphine HCl	2	Stable	17
Novobiocin	0	Degrades	10
	4	Stable	10
Neomycin	0	Degrades	10
	4	Stable	10
Dihydrostreptomycin	0	Degrades	10
	4	Stable	10
Heparin	0	Degrades	19
	1	Degrades	19
Mitomycin	5	Stable	20
Bleomycin HCl	5	Stable	20
5-Fluorouracil	5	Stable	20
Aspirin	5	Stable	20
Sulfanilamide	5	Stable	20
Salicylic acid	5	Stable	20
Colchicine	5	Stable	20
Nafarelin acetate	6	Stable	21
[D-Trp[6]]LH-RH	6	Stable	22

[a]   0, dry powder in air; 1, aqueous solution; 2, ointment base; 3, impregnated paper strips; 4, peanut oil base; 5, polymerized in polymer at $-78°C$; and 6, poly(D,L-lactide-co-glycolide) microspheres.

is potentiated by oxygen. Thus, radiation sterilization should be done under vacuum to reduce radiation-induced oxidation in susceptible drugs. The radiation sterilization effects on numerous drugs have been studied and reported. The results are summarized in Table 6.

IV. STERILIZATION EFFECTS ON SELECTED BIOMEDICAL POLYMERS AND MATERIALS

Several papers have been published discussing the general effects of sterilization processes on materials, particularly polymers.[1,2,4,13,19,23-25] Suggested sterilization methods on selected biomedical polymers and materials are summarized in Table 7.

Ionizing radiation effects on polymers have been studied extensively. In general, radiation degrades polymers by (1) chain scission, a random rupturing of bonds which reduces the molecular weight of the polymer, and (2) cross-linking of the polymer molecules, resulting in the formation of large three-dimensional networks. As a result of chain scission, low

Table 7
SUITABLE STERILIZATION METHODS FOR SELECTED BIOMEDICAL POLYMERS AND MATERIALS[2,4,17,23,24]

Material	Suitable methods[a]	Comments
Acetals	2, 5, 6	Avoid phenolic or acidic agents; steam heat marginal for polyformaldehyde
Acrylics	3, 5, 6	Radiation marginal; polymethylmethacrylate will tolerate one sterilizing radiation; polyacrylonitrile, polyacrylates, polyhydroxyacrylates, and polycyanoacrylates tolerate radiation better than polymethylmethacrylate
Polyamides	2, 3, 5, 6	Avoid phenolic agents; some grades have high water absorption; aramids have excellent radiation stability.
Vinyls	3, 5, 6	Steam heat marginal on flexible polyvinyl chloride; avoid phenolic agents; ethylene oxide marginal on both rigid and flexible polyvinyl chloride; do not apply ethylene oxide to previously irradiated items
Styrene and copolymers	3, 5, 6	Poly(styrene-acrylonitrile) crazed by excess ethylene oxide; polystyrene the most radiation stable of thermoplastics; acrylonitrile-butadiene-styrene will tolerate at least one sterilizing radiation dose
Polyolefins		
Polyethylene	3, 5, 6	Wetting problems with liquid chemicals; some
Polypropylene	2, 5, 6	copolymers of polypropylene are stable to
Polymethylpentane	1, 2, 3, 5, 6	sterilizing radiation
Ethylene vinyl acetate	3, 5, 6	
Fluorocarbon polyers		
Polytetrafluoroethylene	1, 2, 5, 6	Polytetrafluoroethylene extremely sensitive to
Polytrifluorochloroethylene	1, 2, 3, 5, 6	radiation
Fluorinated ethylene/propylene resins	1, 2, 3, 5, 6	
Cellulosics	3, 5, 6	Some cellulosics soluble in liquid chemicals; radiation may reduce strength
Polysulfones	1, 2, 3, 5, 6	
Phenol, melamine, or urea formaldehyde resins	1, 2, 3, 5, 6	Not resistant to acids; urea and melamine formaldehyde resins not stable to dry heat
Polycarbonates	2, 3, 5, 6	Steam heat marginal; polycarbonates not resistant to alkali; limited resistance to detergents
Polyesters	2, 3, 5, 6	
Epoxy resins	2, 3, 5, 6	Some resins and catalysts may be steam sterilized separately
Polyimides	1, 2, 3,	
Polyurethanes	3, 5, 6	Thermal stability variable; avoid acids or alkalis
Silicones		
Fluids	1, 2, 4	Liquid chemicals marginal
Rubbers	1, 2, 3, 5, 6	
Natural rubbers	2, 3	
Gelatin	1, 4	
Polyvinyl pyrrolidone	2, 3	
Glass and metals	1, 2, 3	

[a]   1, dry heat; 2, steam heat; 3, ionizing radiation; 4, microfiltration; 5, ethylene oxide; and 6, liquid chemical.

molecular weight fragments, gas evolution (generally hydrogen), and unsaturated bonds may appear. In addition, crystallinity, hence density, may also change. Cross-linking generally results in an initial increase in tensile strength and a decrease in impact strength, the polymer becoming more brittle with increased radiation dose. Upon irradiation of polymers, color may develop, influenced by oxygen upon storage. Oxygen may also influence the effects of radiation on polymers as free radicals produced in the polymers may convert to peroxides.[24]

In general, vinyl polymers such as polyvinyl chloride, polyethylene, polypropylene, and polystyrene are cross-linked by radiation. Vinylidene polymers, such as polyvinylidene chloride and polymethacrylate, and cellulosic polymers degrade when irradiated. Addition polymers, such as polyesters and polyamides, cross-link upon irradiation. Aromatic groups decrease the amount of degradation (both cross-linking and chain scission) caused by radiation. Aromatic plasticizers, free radical scavengers, and most antioxidants decrease both cross-linking and chain scission degradation. Molecular oxygen interferes with cross-linking, but may enhance chain scission degradation.

Numerous reports have focused on defining the proper sterilization method for, or the effects on, specific or special materials. de Koning et al.[26] have developed a technique for cross-linking agarose-encapsulated sorbents that are resistant to steam sterilization at 122°C for at least 2 hr and at 134°C for at least 30 min. When radiation sterilized, the cross-linked agarose beads degraded. No significant influence of oxygen was observed. Ethylene oxide sterilization was not a viable alternative as adsorption to the encapsulated sorbents can be a problem. The degradative effects of steam sterilization on pure polycrystalline germanium and a cast surgical vitallium Co/Cr/Mo alloy were addressed by Baier et al.[27] The results of this study suggested that the properties of carefully prepared biomedical implants could be compromised by the deposition of hydrophobic organic and hygroscopic salt contaminants on the implant surface via steam sterilization. In general, exceeding the glass transition temperature of polymers using dry heat or steam sterilization induces morphological and surface changes in the polymer.[1]

White and Bradley studied the ethylene oxide residues in medical-grade silicones and Dacron®-reinforced silicone sheeting.[28] Their study showed that diffusion of ethylene oxide in the medical-grade silicones is extremely rapid so that very rapid desorption and low residue levels are observed. In contrast, Dacron®-reinforced silicone sheeting desorbed ethylene oxide much slower due to retention by the Dacron®. In other ethylene oxide residue studies, Gilding et al.[29] studied the effect of hydrophobic/hydrophilic balance, crystallinity, charge, and plasticization on the ethylene oxide residues in crystalline polymers, glassy polymers, and elastomers. Ball determined residues of ethylene oxide and its two by-products, ethylene glycol and ethylene chlorohydrin, in aqueous ophthalmic solutions.[30] A mathematical model has been developed to evaluate ethylene oxide desorption from ethylene oxide-sterilized materials under different physicochemical conditions as well as in vivo.[31] In another study, Handlos studied ethylene chlorohydrin formation in radiation and ethylene oxide-sterilized polyvinyl chloride.[32] The results of this study indicated that the ethylene chlorohydrin content depended on the heat aging of the polymer during manufacturing and the time lapse between irradiation and subsequent ethylene oxide treatment.

Ethylene oxide sterilization of plastic tubing used for extracorporeal bypass induces hemolysis of blood.[33] Ethylene oxide-sterilized tubing should aerate for at least 5 days before use in extracorporeal bypass. The interaction of ethylene oxide with plasticized polyvinyl chloride has been studied.[34] Based upon the solubility and partial vapor pressure of ethylene oxide in a homologous series of phthalic acid esters, the structure and concentration of plasticizers played an important role in determining the most effective degassing procedure. As the length of the hydrocarbon ester moiety decreased, the ethylene oxide solubility increased, necessitating increased degassing times. The partial vapor pressure of ethylene oxide decreased as a function of the length of the dialkyl chain, suggesting that the lower esters would degas more rapidly. Because of the interplay between ethylene oxide solubility and partial vapor pressure, O'Leary and Guess rationalized that the higher alkyl phthalate esters would be the preferred plasticizers for ethylene oxide-sterilized plastics.[34] In the same study,[34] the hemolytic liability of 17 pharmaceutical plastics was addressed. The results indicated that polypropylene was the best candidate material for making blood-contacting medical devices sterilized by ethylene oxide.

The effect of ethylene oxide and ^{60}Co sterilization on the biodegradable copolymers composed of poly(ethylene oxide)/poly(ethylene terephthalate) has been studied.[35] Dry heat at 100°C for 16 hr under vacuum showed no reduction on the molecular weight of the copolymers. Ethylene oxide sterilization at 55°C showed no decrease in molecular weight. ^{60}Co irradiation at 2.5 Mrad caused a 50% reduction in molecular weight. Since these copolymers undergo degradation by hydrolysis (cleavage at the ester bond linking the ester and hydrophilic ether segments), a hydrolytically unstable copolymer is created by ^{60}Co sterilization. Similar reductions in molecular weight due to ^{60}Co irradiation have been noted with other biodegradable materials. ^{60}Co sterilization at doses of 2.5 Mrad is known to cause deterioration of Dexon® and Vicryl® (polyglycolic acid) sutures.[36,37] Large quantities of low molecular weight material are formed by unzipping, the sequential elimination of monomer units from the main polymer chain ends (chain depolymerization). Thus, the initial strength of polyglycolic acid sutures maintained by the crystalline regions is unchanged by radiation sterilization, but upon hydrolysis the tensile strength falls to zero after 10 days of implantation. The hydrolytic degradation rate of ^{60}Co-irradiated polymer is greater than unirradiated polymer so that catastrophic mechanical failure occurs in days rather than weeks. A decrease in intrinsic viscosity (thus molecular weight) has also been noted in poly(D,L-lactide-co-glycolide) microspheres upon exposure to gamma irradiation[21,38] A decrease in the intrinsic viscosity with an increase in radiation dose was noted. The magnitude of decrease in the intrinsic viscosity was independent of the lactide/glycolide mole ratio.[38]

The suitability of using gamma radiation as an alternative sterilization method to dry heat sterilization for polyethylene glycols was studied by Bhalla et al.[39] The viscosity, freezing point, hydroxyl value, average molecular weight, ethylene glycol, and diethylene glycol content of the polyethylene glycols were not significantly altered by either method of sterilization. In contrast to dry heat-sterilized polyethylene glycols, the irradiated polyethylene glycol samples showed no color change.

V. RELEVANCE TO CONTROLLED RELEASE SYSTEMS

Of the controlled release systems available today, the majority of them are diffusion-controlled polymeric systems governred by Fick's first law of diffusion:

$$J = -D \frac{dC}{dx} \tag{3}$$

where $J = dM/dt$, which is the flux (mass of drug) across a plane surface of unit area per unit time; D = diffusivity of the drug molecule in the diffusion medium; and dC/dx = the concentration gradient of the drug across a diffusion path with thickness of dx. The negative sign is used to demonstrate that the direction of diffusion is in the direction of decreasing drug concentration. When D is independent of x, Fick's second law can be derived from Fick's first law.[40] In three dimensions,

$$\frac{dC}{dt} = D \left[\frac{d^2C}{dx^2} + \frac{d^2C}{dy^2} + \frac{d^2C}{dz^2} \right] \tag{4}$$

Fick's second law implies that the rate of change in drug concentration in a volume element within the diffusional field is proportional to the rate of change in concentration gradient at that point in the field. The proportionality constant is the diffusivity, D. For the unidimensional diffusion case, Equation 4 simplifies, for example, to:

$$\frac{dC}{dt} = D\frac{d^2C}{dx^2} \tag{5}$$

Both Fick's first and second laws of diffusion can be used to develop mathematical expressions for the controlled release of drugs from capsule- (membrane-bound reservoir) and matrix- (monolithic) type controlled delivery systems. For the capsule-type controlled release system, the rate of drug release can be defined by:[41]

$$\frac{Q}{t} = \frac{C_p K D_s D_p}{K D_s h + D_p S_d} \tag{6}$$

where Q = cumulative amount of drug released per unit surface area; t = time; C_p = solubility of drug in the polymer (membrane); $K = C_s/C_p$, where C_s = drug solubility in environmental (aqueous) solution; D_s = drug diffusivity in environmental solution; D_p = drug diffusivity in polymer (membrane); h = thickness of the membrane; and S_d = thickness of the hydrodynamic boundary diffusion layer. For the matrix-type controlled release system, the cumulative amount of drug released from a thickness of drug dispersion zone, h, can be defined by: [41]

$$Q = \left(A - \frac{C_p}{2}\right)h \tag{7}$$

where A = the initial drug loading in the polymer matrix. The magnitude of h is a function of time, dependent upon the physicochemical interactions between drug molecules and the polymer matrix, and can be expressed by:[41]

$$h^2 + \frac{2(A - C_p)D_p h S_d}{\left(A - \frac{C_p}{2}\right)D_s kK} = \frac{2C_p D_p}{\left(A - \frac{C_p}{2}\right)} \tag{8}$$

where k = constant accounting for the relative magnitude of the concentration gradients in both h and S_d.

It can be seen from Equations 6 to 8 that the rate-controlling physicochemical parameters for the controlled release of drugs from both capsule-and matrix-type controlled release systems are the drug solubilities, C_p and C_s, diffusivities, D_p and D_s, the partition coefficient, K, and the diffusional path thicknesses, h and S_d. Thus, for a controlled release system to function reliably and reproducibly, the physicochemical parameters need to be controlled and well defined. Changes in the physicochemical parameters due to processing, aging, or the effects of sterilization need to be evaluated in light of the effects on the drug release kinetics and ultimate drug bioavailability, efficacy, or toxicity.

Perhaps the two major determinants of the rates of drug release from either the capsule- or matrix-type drug delivery systems are the polymer solubility of the drug, C_p, and the diffusivity in the polymer, D_p. The difference in polymer solubilities among drugs is large, illustrated by the wide range of solubilities exhibited by the steroids in liquid silicone polymer.[41] Stereochemical configurations and variation in functional groups, as well as the degree of drug hydration, may affect the magnitude of drug polymer solubility. We have seen that ethylene oxide is very reactive with a number of chemicals. Thus, the potential exists for changing functional group stereochemistry of many drugs exposed to ethylene oxide sterilization, changing the drug polymer solubility. Even if the drug does not react with ethylene oxide, if sufficient reactive water is available, ethylene glycol may form, acting as a solubilizing agent for the drug as well as a plasticizer for the polymer. In either case, the inherent polymer solubility of the drug is changed, resulting in potentially undesirable drug release kinetics. Ionizing radiation also can have direct effects on drugs, altering the drug solubility in the polymer. However, the more likely phenomenon is the effect of the radiolytic products of water upon the drug. Many drugs, particularly proteins, are sensitive to oxidation and thus are susceptible to dry heat, UV radiation, and ionizing radiation sterilizations. For example, it has been reported that ionizing radiation causes degradation of steroids by oxidation at the C-11 position and oxidative cleavage of the corticoid side chain at the C-17 position.[10] Even if the potency of the drug is not compromised, significant alteration of the drug may occur to change its polymer solubility.

Ultimately, any variation in the polymer solubility of the drug, even if independent of a change in the aqueous drug solubility, C_s, will result in an increase or decrease in the partition coefficient, K. This becomes important in matrix-type controlled release systems where drug release can be a partition-controlled process or a matrix-controlled process dependent upon the magnitude of the partition coefficient. The partition coefficient also varies dependent upon the type of biomedical polymer,[41] so that sterilization-induced material changes may also affect the partition coefficient and ultimately the release kinetics.

The drug diffusivity in the polymer is governed by many parameters. From the hole theory of diffusion,[42] the rate of diffusion will depend on (1) the number and size distribution of preexisting holes and (2) the ease of hole formation. The number and size distribution of preexisting holes depends upon the degree of packing of the polymeric chains, related to the free volume and density. The ease of hole formation depends upon the segmental chain mobility (chain stiffness) and the cohesive energy of the polymer. The coefficient of thermal expansion and the glass transition temperature reflect these features. In addition, the magnitude of the drug diffusivity in the polymer is dependent upon the types of functional groups, molecular weight and stereochemical configuration of the drug molecule, the degree of polymer crystallinity, and cross-linking, as well as additives such as fillers and plasticizers.[41,42] An increase in the polymer cross-linking, induced by radiation sterilization, reducing the polymer chain mobility, results in a reduction in polymeric drug diffusivity. For example, in the release of Norgestomet from a hydrogel matrix, the reduction in polymeric drug diffusivity was a linear function of the reciprocal of the degree of cross-linking.[41] Crystallinity acts similarly to cross-linking by introducing regions of relatively low diffusion in contrast to the surrounding amorphous structure, resulting in a reduction in the gross drug diffusivity.[41] Crystallinity induced by ionizing radiation or elevated temperatures above the glass transition temperature of some polymers results in decreased polymeric drug diffusivity and potentially deleterious drug release kinetics.

The effects of sterilization of the rate-controlling physicochemical parameters of controlled release systems are varied and may not be immediately apparent. Many of the changes induced by sterilization processes are time-dependent phenomena, so that only upon storage and aging do changes in the controlled release system become evident. Thus, in order to fully evaluate sterilized controlled release systems, it is imperative that the appropriate stability study guidelines be developed, addressing both drug potency and drug release kinetics.

ACKNOWLEDGMENTS

I would like to thank Dr. M. Akers for his helpful comments and Mrs. Bonita Brooks and Mr. David Moore for assistance in preparation of the manuscript.

REFERENCES

1. **Bruck, S. D.,** Sterilization problems of synthetic biocompatible materials, *J. Biomed. Mater. Res.,* 5, 139, 1971.
2. **Plester, D. W.,** The effects of sterilising processes on plastics, *Bio-Med. Eng.,* p. 443, September 1970.
3. *The United States Pharmacopeia,* 20th revision, U. S. Pharmacopeial Convention, Inc., Rockville, Md., 1980, 1037.
4. **Bloch, B. and Hastings, G. W.,** *Plastics Materials in Surgery,* 2nd ed., Charles C Thomas, Springfield, Ill., 1972, chap. 6.
5. **Abshire, R. L.,** High-intensity ultraviolet radiation to sterilize empty polyethylene bottles, *Pharm. Manuf.,* p. 16, January 1985.
6. **Myrvik, Q. N., Pearsall, N. N., and Weiser, R. S.,** *Fundamentals of Medical Bacteriology and Mycology,* Lea & Febiger, Philadelphia, 1978, 65.
7. **Williams, D. F. and Roaf, R.,** *Implants in Surgery,* W. B. Saunders, London, 1973, 246.
8. **Morrissey, R. F.,** An Introduction to Radiation Technology, paper presented at Pharmaceutical Manufacturers Association Seminar on Radiation Sterilization, Atlanta, November 30 to December 2, 1981, 6.
9. **Heid, J. L. and Joslyn, M. A.,** *Fundamentals of Food Processing Operations,* AVI Publishing, Westport, Conn., 1967, 333.
10. **Tsuji, K., Kane, M. P., Rahn, P. D., and Steindler, K. A.,** ^{60}Co irradiation for sterilization of veterinary mastitis products containing antibiotics and steroids, *Radiat. Phys. Chem.,* 18(3-4), 583, 1981.
11. **Rendell-Baker, L. and Roberts, R. B.,** Gas versus steam sterilization: when to use which, *Med. Surg. Rev.,* 4th quarter, 10, 1969.
12. *The Merck Index,* 10th ed., Merck and Co., Inc., Rahway, N.J., 1983.
13. **Yasuda, H., Refojo, M. F., and Stone, W.,** Sterilization of polymers, *Trans. Am. Chem. Soc.,* p. 209, September 1964.
14. **DeRisio, R. J.,** Radiation Sterilization: Regulatory Status (U.S.), paper presented at Pharmaceutical Manufacturers Association Seminar on Radiation Sterilization, Atlanta, November 30 to December 2, 1981, 40.
15. *Fed. Regist.,* 43(122), 4110-03, June 23, 1978, 27474.
16. *Fed. Regist.,* 43(19), 6560-01, January 27, 1978, 3801.
17. **Gopal, N. G. S., Rajagopalan, S., and Sharma, G.,** Feasibility Studies on Radiation Sterilization of Some Pharmaceutical Products, Paper IAEA/SM-192/6, in Proc. Symp. Ionizing Radiation for Sterilization of Medical Products and Biological Tissues, International Atomic Energy Agency, Bombay, 1974.
18. **Jacobs, G. P.,** The Effect of Radiation on Drugs, paper presented at Pharmaceutical Manufacturers Association Seminar on Radiation Sterilization, Atlanta, November 30 to December 2, 1981, 139.
19. **Blackburn, R., Iddon, B., Moore, J. S., Phillips, G. O., Power, D. M., and Woodward, T. W.,** Radiation Sterlization of Pharmaceuticals and Biomedical Products, Paper IAEA/SM-192/16, in Proc. Symp. Ionizing Radiation for Sterilization of Medical Products and Biological Tissues, International Atomic Energy Agency, Bombay, 1974.
20. **Kaetsu, I., Yoshida, M., and Yamada, A.,** Controlled slow release of chemotherapeutic drugs for cancer from matrices prepared by radiation polymerization at low temperatures, *J. Biomed. Mater. Res.,* 14, 185, 1980.
21. **Sanders, L.M., Kent, J. S., McRae, G. I., Vickery, B. H., Tice, T. R., and Lewis, D. H.,** Controlled release of a luteinizing hormone-releasing hormone analogue from poly-(d,1-lactide-co-glycolide) microspheres, *J. Pharm. Sci.,* 73(9), 1294, 1984.
22. **Redding, T. W., Schally, A. V., Tice, T. R., and Meyers, W. E.,** Long-acting delivery systems for peptides: inhibition of rat prostate tumors by controlled release of [D-Trp^6] luteinizing hormone-releasing hormone from injectable microcapsules, *Proc. Natl. Acad. Sci. U.S.A.,* 81, 5845, 1984.
23. **Plester, D. W.,** The effects of radiation sterilization on plastics, in *Industrial Sterilization,* Phillips, C. B. and Miller, W. S., Eds., Duke University Press, Durham, 1973, chap. 10.

24. **Skiens, W. E. and Williams, J. L.,** Ionizing Radiation's Effects on Selected Biomedical Polymers, paper presented at Radiation Sterilizers Incorporated Seminar on Radiation Sterilization, Schaumburg, October 26, 1982, 1.

25. **Williams, J. L. and Dunn, T. S.,** Material Effects During Radiation Sterilization, paper presented at Pharmaceutical Manufacturers Association Seminar on Radiation Sterilization, Atlanta, November 30 to December 2, 1981, 130.

26. **de Koning, H. W. M., Chamuleau, R. A. F. M., and Bantjes, A.,** Crosslinked agarose encapsulated sorbents resistant to steam sterilization. Preparation and mechanical properties, *J. Biomed. Res.*, 18, 1, 1984.

27. **Baier, R. E., Meyer, A. E., Akers, C. K., Natiella, J. R., Meenaghan, M., and Carter, J. M.,** Degradative effects of conventional steam sterilization on biomaterial surfaces, *Biomaterials,* 3, 241, 1982.

28. **White, J. D. and Bradley, T. J.,** Residual ethylene oxide in gas-sterilized medical-grade silicones, *J. Pharm. Sci.,* 62(10), 1634, 1973.

29. **Gilding, D. K., Reed, A. M., and Baskett, S. A.,** Ethylene oxide sterilization: effect of polymer structure and sterilization conditions on residue levels, *Biomaterials,* 1, 145, 1980.

30. **Ball, N. A.,** Determination of ethylene oxide, ethylene chlorohydrin, and ethylene glycol in aqueous solutions and ethylene oxide residues in associated plastics, *J. Pharm. Sci.,* 73(9), 1305, 1984.

31. **Handlos, V.,** Kinetics of the aeration of ethylene oxide sterilized plastics, *Biomaterials,* 1, 149, 1980.

32. **Handlos, V.,** Ethylene chlorohydrin formation in radiation- and ethylene oxide-sterilized poly(vinyl chloride), *Biomaterials,* 5, 86, 1984.

33. **Hirose, T., Goldstein, R., and Bailey, C. P.,** Hemolysis of blood due to exposure to different types of plastic tubing and the influence of ethylene-oxide sterilization, *J. Thorac. Cardiovasc. Surg.,* 45(2), 245, 1963.

34. **O'Leary, R. K. and Guess, W. L.,** The toxicogenic potential of medical plastics sterilized with ethylene oxide vapors, *J. Biomed. Mater. Res.,* 2, 297, 1968.

35. **Reed, A. M. and Gilding, D. K.,** Biodegradable polymers for use in surgery — poly(ethylene oxide)/poly(ethylene terephthalate) (PEO/PET) copolymers. II. *In vitro* degradation, *Polymer,* 22, 499, 1981.

36. **Gilding, D. K. and Reed, A. M.,** Biodegradable polymers for use in surgery — polyglycolic/poly(lactic acid) homo- and copolymers. I, *Polymer,* 20, 1459, 1979.

37. **Chu, C. C. and Williams, D. F.,** The effect of gamma irradiation on the enzymatic degradation of polyglycolic acid absorbable sutures, *J. Biomed. Mater. Res.,* 17, 1029, 1983.

38. **Beck, L. R. and Tice, T. R.,** Poly(lactic acid) and poly(lactic acid-co-glycolic acid) contraceptive delivery systems, in *Long-Acting Steroid Contraception,* Mishell, D. R., Ed., Raven Press, New York, 1983, 175.

39. **Bhalla, H. L., Menon, M. R., and Gopal, N. G. S.,** Radiation sterilization of polyethylene glycols, *Int. J. Pharm.,* 17, 351, 1983.

40. **Crank, J.,** *The Mathematics of Diffusion,* Oxford University Press, Oxford, 1975, chap. 1.

41. **Chien, Y. W.,** *Novel Drug Delivery Systems,* Marcel Dekker, New York, 1982, chap. 9.

42. **Crank, J. and Park, G. S.,** *Diffusion in Polymers,* Academic Press, New York, 1968, chaps. 2, 3, and 6.

INDEX

R

R-2323, 134
Radiation of microorganisms, 167
Radiation sterilization, 165—167, 171, 174
Radioactive isotopes, 146
Radioactive tag, 9
Radioimmunoassay, 149
Radiopaque materials, 9
Radiotherapy, 152
Radiotracing, 34
Reconstituted collagin, 87
Redox drug delivery, 158
Reservoir, bioerodible, 84
Resins, 46
Rheology, 31—33
Riboflavin-5´-sodium phosphate, 6
Ricin, 151—152
Roller feeds, 121
Room-temperature-vulcanizing (RTV) systems, 112,
    117, 126, 128, 133, 135—136
RTA, 151—153
RTB, 151—152
RTV, see Room-temperature-vulcanizing

S

SA, see Sebaic acid
Salicylic acid, 27, 36, 52
Salt precipitation, 149
Scale-up problems, 4—5
Scanning electron microscopy (SEM), 6
Schistosoma antigens, 17
Screening assay, 149
Screw design, 121
Sebaic acid (SA), 98—99
SEM, see Scanning electron microscopy
[^{75}Se] norcholestenol, 8
Sesame oil, 29, 88
Silica gel, 49, 51
Silicon dioxide gel, 50
Silicone oil, 67
Silicone rubber fabrication, 111—129
  high-consistency, 112—125
    calendering, 124
    extruding, 120—123
    molding, 120
    postcuring, 124—126
    preparation, 115—117
    vulcanizing, 117—120, 123—124
      agents for, 117—118
      amount of agent for, 118
      in autoclaves, 125
      effect of temperature and agent on, 119
      hot-air, 123—124
      methods and equipment for, 123—124
      problems of, 118—119
      of thick sections, 119—120
  one-part low-consistency, 128
  two-part low-consistency, 126—128
    coextrusion, 127—128

    extruding, 127
    vulcanizing, 126—127
Silicone rubbers, 112—114, 126
Site-Release® System, 54
Sodium alginate
  antigen mixed with, 147
  cross-linking of, 45
  effects on nicotinic acid tablets, 56
  gel formation by, 44, 48
  granules made from, 53
  incorporation of ketoprofen into, 55
  sustained-release action of, 51
Sodium carbonate, 102—103
Sodium carboxymethylcellulose, 48, 51—52, 57
Sodium chloride, 37, 68
Sodium cromoglycate, 13
Sodium hydroxide, 52, 68
Sodium metabisulfite, 13
Sodium salicylate, 56—57
Sodium silicate, 49
Sodium sulfate, 103
Sodium sulfite, 13
Solid-phase radioimmunoassay, 149
Sorbitan mono-oleate, 26, 30, 35
Sorbitol, 37, 44
Soybean globulin, 50
SPDP, see Succinimidyl-3-(2-
    pyridyl[dithio]propionate)
Spermicidal agents, 137
Spherex®, 17
Spleen cells, 147—148
Spleen embolization, 157
Squalane, 29
Starch vegetable gums, 42
Starch xanthan granules, 53
Stearic acid, 51
Stearyl alcohol, 51
Sterilization of controlled release systems, 163—176
  effect on drugs, 169—171
  effect on selected biomedical polymers and
    materials, 171—174
  methods for, 165—169
  relevance of, 174—176
Streptococcus mutans, 56
Streptozotocin, 11
Streptozotocin-diabetic rats, 9
Styrene, 63—64, 68
Succinimidyl-3-(2-pyridyl[dithio]propionate)
    (SPDP), 153
Sucrose, 33
Sulfabenzamide, 52
Sulfacetamide, 52
Sulfadiazine, 93
Sulfathiazole, 52
Sulfuric acid, 49
Surfactant I, 24, 27—28, 35
Surfactant II, 27—28, 35
Surfactant selection, 29—31
Suspension polymerization, 4
Sustained release products, 46—48, 56
Swelling-controlled mechanism, 73—75